APPLICATIONS OF MODIFIED STARCHES

Edited by **Emmanuel Flores Huicochea**
and **Rodolfo Rendón Villalobos**

Applications of Modified Starches
http://dx.doi.org/10.5772/intechopen.68610
Edited by Emmanuel Flores Huicochea and Rodolfo Rendón Villalobos

Contributors

Aurelio Ramirez Hernández, Jose Luis Rivera-Armenta, Maria Leonor Mendez-Hernandez, Zahida Sandoval-Arellano, Beatriz Adriana Salazar- Cruz, Maria Yolanda Chavez-Cinco, Monica Roxana Nemtanu, Mirela Brasoveanu, Farayde Matta Fakhouri, Normane Chaves, Rozanna Fialho, Elaine Albuquerque, Fernando Freitas De Lima, Emmanuel Flores Huicochea

Notice

First published in London, United Kingdom, 2018 by IntechOpen
IntechOpen is the global imprint of INTECHOPEN LIMITED, registered in England and Wales, registration number: 11086078, The Shard, 25th floor, 32 London Bridge Street
London, SE19SG – United Kingdom
Printed in Croatia

British Library Cataloguing-in-Publication Data
A catalogue record for this book is available from the British Library

Additional hard copies can be obtained from orders@intechopen.com

Applications of Modified Starches, Edited by Emmanuel Flores Huicochea and Rodolfo Rendón Villalobos
p. cm.
Print ISBN 978-1-78923-372-8
Online ISBN 978-1-78923-373-5

For EU product safety concerns:
IN TECH d.o.o., Prolaz Marije Krucifikse Kozulić 3, 51000 Rijeka, Croatia,
info@intechopen.com or visit our website at intechopen.com.

Meet the editors

Ph. D. Emmanuel Flores Huicochea, PhD, obtained his bachelor and MSc degrees in Chemical Engineering from Universidad Autónoma del Estado de Morelos, México and later obtained a PhD in Engineering degree from the Chemistry Faculty, Universidad Nacional Autónoma de México, in 2013. He is a full-time researcher and associate professor of postgraduate in the Centro de Desarrollo de Productos Bióticos belonging to Instituto Politécnico Nacional, México; additionally he has contributed in research projects and has published research papers and book chapters. His investigation topics are related to starch modification and use in order to produce bioplastics added with lignocellulosic to use like barrier or disposable container.

Rodolfo Rendón Villalobos is a research professor at Instituto Politécnico Nacional, Mexico. He obtained his academic degree in Biology from the Autonomous University of Morelos, master's degree in Marine Ecology from CICESE and PhD degree in Polymer Science from Zacatepec Institute of Technology. He has honors and awards; such as National System of Researches (SNI) fellow, Associazione Italiana di Scienza e Tecnologia delle Macromolecole fellow and International Association for Cereal Science and Technology fellow. He has more than 50 scientific publications, 6 book chapters, over 100 participations in congresses and scientific conferences. Professor Rendón has expertise in chemical modifications as well as in development of biopolymers; evaluation in structural characteristics as FT-IR, SEM, XRD; and also biodegradation studies and rheological and thermal analysis.

Contents

Preface

Intech Open editorial offered us the opportunity to be editors.The topic that matched into our expertise was "Starch modifications" and we began our journey of being editors. This book represents the final results and it is our first book on this theme.

The book describes starch modifications from physical treatments to ionizing radiation process, the key properties of modified starch, and its application fields.

The book is composed of six chapters and the main purpose of the book is provide to the readers a view of modification procedures widely used through starch, description of the main characteristics about modified starches and finally it is providing application fields and application examples. The first chapter provides an overview of the procedures for modifying starches commonly used. The second chapter analyzes the graft process using synthetic polymer and starch from point of view of reaction conditions and solvent type use during the reaction. The use of acetylate starch on pharmaceutical industry is analyzed in depth.

The third chapter shows us evaluation of styrene content over physical and chemical properties of elastomer/TPS-EVOH/chicken feather composites.

The fourth chapter describes production and characterization of starch nanoparticles. The last chapter focuses on an unusual procedure to modify starch using radiation that comes from several sources like UV, gamma or accelerated electrons. Throughout the chapter, the readers could know reaction conditions and characteristics of starches attained.

We hope the book will be useful.

Emmanuel Flores Huicochea
National Polytechnic Institute
Yautepec, Morelos, Mexico

Rodolfo Rendón Villalobos
National Polytechnic Institute
Center for Development of Biotic Products
Yautepec, Morelos, Mexico

Chapter 1

Introductory Chapter: Starch Modifications

Emmanuel Flores Huicochea

Additional information is available at the end of the chapter

http://dx.doi.org/10.5772/intechopen.78366

1. Overview of modifications

Starch is the main polysaccharide used as food and ingredient or additive in the food industry. The main advantages over other polysaccharides are the low cost and renewable raw material, but even starch can be obtained from several natural sources or modified by engineering genetic, as the waxy starch of corn. It also has drawbacks such as: lower solubility usually in water, lower shear stress resistance, lower capacity to support thermal decomposition, and high retrogradation, and these issues limit the use on several temperatures and pH usually found on food system and industrial applications.

The native starch can be improved by means of physical, chemical, and ionic treatments. The physical treatments are cost-effective, are less labor intensive, and only have initial investment of equipment than the chemical treatments (**Figure 1**). The aim of physical treatment is to produce changes in the granular structure, and the common processes involved are heat moisture, annealing, and extrusion. However, today, there are novel processes like sonication and high hydraulic pressure. **Figure 2** shows the main characteristics of these processes. This plot is an easy way to view and compare process conditions like moisture, temperature (glass and gelatinization), and the effect of treatment on granule structure and molecular weight. For example, after annealing process, the granular structure and molecular weight remains without change, and the moisture percent is similar that heat moisture processes employ, but the temperature of process is set between glass and gelatinization temperature.

The chemical modified starch has lot of reports in the literature compared with the other treatments; on this spectrum of chemical modifications, the starch molecules can reduce the molecular weight of chains, add functional groups like acetyl, or even oxidize OH groups present in the monomer of the starch polymers. The chemical treatments also can associate chains of starch like in cross-linked reaction or even make a web around the starch granule. Another step forward in the reactions, complexity of starch is graft copolymerization, which means adding synthetic monomers or polymers or semisynthetic polymers to the biopolymers of starch. Another interesting

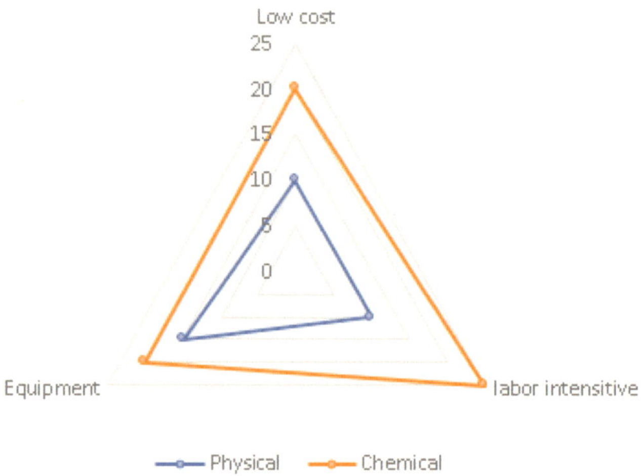

Figure 1. Relative comparison of the physical and chemical treatments of starch.

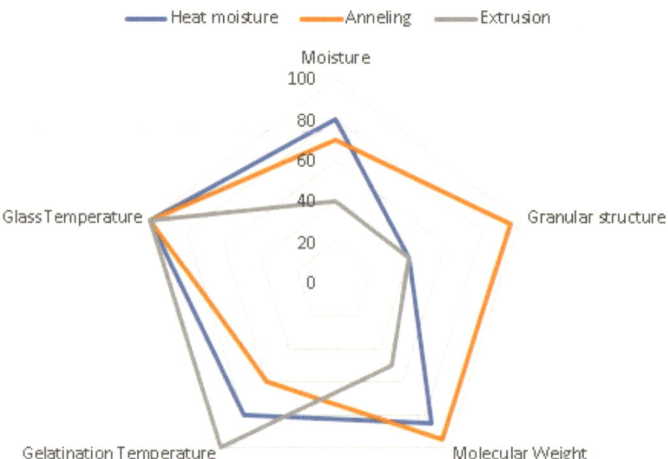

Figure 2. Relative comparison of process conditions and its effect on granular structure for physical treatments of starch.

process to modify the starch is the ionizing radiation process. The radiation ionizing can come from UV, gamma, or accelerated electrons; at first instance, this process was used with synthetic polymers, but today it is used with starch. The radiation produces free radicals that react with starch polymers to produce different properties on the ionized polymers.

Author details

Emmanuel Flores Huicochea

Address all correspondence to: efloreshu@outlook.com

Instituto Politécnico Nacional, Centro de Desarrollo de Productos Bióticos, Yautepec, Morelos, Mexico

Chapter 2

Chemical Modification of Starch with Synthetic

Aurelio Ramírez Hernández

Additional information is available at the end of the chapter

http://dx.doi.org/10.5772/intechopen.72384

Abstract

An alternative for solving the environmental pollution problems generated by conventional plastics, it is the chemical modifications graft-type of the starch with synthetic polymers of post-consumer and in situ polymerizations on the starch granules. The starch modified by this methodology allows to counteract the disadvantages of both polymers such as the little or no biodegradability of the synthetic polymer and the poor mechanical properties of the starch. In the present study, a review on the chemical modification of starch with synthetic polymers by grafting is carried out. Factors affecting the copolymerization reactions of starch-g-synthetic polymer were analyzed, for example, their chemical nature, solubility, size and length of polymer chains, temperature, catalyst and starch/amylose content, as well as their characterization chemistry and the potentials applications of this copolymer.

Keywords: starch, synthetic polymer, graft copolymer, biodegradable

1. Introduction

Due to environmental pollution generated by conventional plastics such as polyethylene (PE), polypropylene (PP), vulcanized rubber, polystyrene (PS) and polyvinyl acetate (PVC), to name a few, several investigations have been carried out to generate materials that can compete or be an alternative to the excessive use of these plastics because these are not biodegradable. From this point of view, the use of natural polymers such as starch can contribute to reduce the negative impact of conventional plastics. However, materials made from starch alone generate plastics, which have important disadvantages such as their high affinity for moisture and poor mechanical properties. An alternative for solving these disadvantages is the chemical modifications of the starch with synthetic polymers of postconsumer and in situ polymerizations on the starch granules. These chemical modifications are called graft-type copolymerizations because the synthetic polymer chains are chemically bonded to the surface of the starch granule, as if these were hair extensions (**Figure 1**).

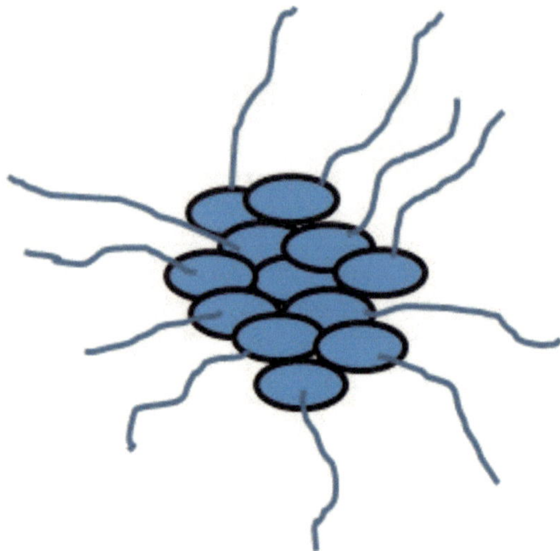

Figure 1. Graft copolymerization of starch-g-synthetic polymer.

Numerous investigations have been reported on physical modifications (i.e., blends or composites) between starch and synthetic polymers [1–5] and few investigations on chemical modifications of copolymerization between these two polymers [6, 7]. This represents a good opportunity to recycle the existing plastic or generate new biodegradable materials through copolymerization reactions. The modified starch by this methodology allows to counteract the disadvantages of both polymers such as the little or no biodegradability of the synthetic polymer and the poor mechanical properties of the starch, to mention some characteristics that generate a wide range of applications of the resulting copolymer. For this reason, the chemical and physical knowledge of synthetic polymer and starch is vital to propose and carry out a possible synthesis of starch-g-synthetic polymer. This chapter analyzes the chemical modification of starch with synthetic polymers by grafting, and factors that affecting the copolymerization reactions of starch-g-synthetic polymer are analyzed, for example, their chemical nature, solubility, size, and so on.

2. Factors affecting the copolymerization reactions of starch-g-synthetic polymer

In order to carry out the chemical modification of the starch with some synthetic polymers, we have to take into account diverse chemical and physical factors. Some of these factors are described below.

2.1. The chemical nature of synthetic polymers

The chemical nature of the polymers is a very important factor to consider for the copolymerization reactions of the starch. The knowledge of the chemical compatibility between the natural and the synthetic polymer, that is, its hydrophilic and hydrophobic nature, will allow to propose a reaction medium to carry out a chemical interaction between these two polymers. For example, polyethylene is a non-polar semi-crystalline polyalkane, which would hardly carry out a copolymerization reaction with the starch by simply mixing both polymers. But the addition of some compatibilizing liquid between PE and starch could facilitate its chemical interaction [1]. Based on the commercial classification of the synthetic polymers and their functional group, the compatibility between the starch and each of the synthetic polymers could be predicted (**Table 1**).

For example, synthetic polymers classified commercially with numbers 2, 4, and 5 would be more difficult to combine with starch compared with classified numbers 1, 3 and 6. On the other hand, there is a great diversity of synthetic polymers classified with number 7 that present chemical compatibility with the starch. The degree of compatibility of the functional groups of the synthetic polymers with the starch has the following order:

Carboxylic acids > acid anhydrides > esters > acid halides > amides > nitriles > aldehydes > ketones > alcohols > mercaptans > amines > ethers > sulfides > alkenes > alkynes > halogenides > nitros > alkanes.

The chemical interaction of the starch with the functional groups of the synthetic polymers results in the production of graft-type copolymers. This type of chemical modification of starch is of great interest worldwide and represents an opportunity to recycle synthetic polymers or to generate materials that are more environmentally friendly and at the same time represent a chemical and physical challenge to improve the properties of starch. Starch graft copolymers are becoming important materials worldwide due to their potential applications in agriculture, medical, and food sectors [8], to mention a few examples.

2.2. Solubility

The solubility of the starch and the synthetic polymer in a given solvent plays a very important role in defining the process or the medium of copolymerization, that is, in homogeneous phase

Synthetic polymer	Commercial classification	Functional group
Poly(ethylene terephthalate), PET	1	Ester
High-density polyethylene, HDPE	2	Alkane, ethyl
Polyvinyl chloride, PVC	3	Vinyl
Low-density polyethylene, LDPE	4	Alkane, ethyl
Polypropylene, PP	5	Alkane, propyl
Polystyrene, PS	6	Aromatic, arylethyl
Others	7*	—

*This number is assigned to more than 100 synthetic polymers.

Table 1. Classification and functional group of synthetic polymers.

or heterogeneous phase [9]. The solubility of a polymer is a function of its chemical nature, chemical structure, molar mass and temperature, to mention some factors that intervene in this. For example, two chains of the same polymer but of different molar mass, the smaller molar mass chain in a given solvent is dissolved more easily than the larger polymer chain. On the other hand, branched polymers are more difficult than their polymer chains to be solvated compared to that of a linear polymer. Natural polymers usually have more complex structures than synthetic polymers; therefore, natural polymers are more difficult to dissolve than synthetic polymers. Polymers having a degree of cross-linking between their chains are more difficult to have their chains solvated than the same polymer but with less or no cross-linking. The structure of the coupling or folding between the different chains of a polymer is important to predict their degree of solubility from the cavities that are generated from these arrangements. On the other hand, the dissolution process of a polymer is different than that of chemical compound or small molar mass. In the case of polymers is a slow process and is composed of two stages or phases. The first stage is called the solvent diffusion process, where the solvent molecules diffuse into the polymer chains producing a swollen gel. The second step is called the decoupling of the polymer chains, that is, the dissolution of the polymer. The latter will occur if the interaction forces between the polymer chains are minor compared to the interaction force of the polymer chains with the solvent. Otherwise, the polymer will only swell, but will not dissolve, as occurs in crystalline and cross-linked polymers [10]. The lighter fractions of a polymer dissolve first, leaving the fractions of high molar mass insoluble. The properties of the polymers in solution are determined by the structural characteristics of the solvated macromolecular chain. The mechanism of dissolution of a polymer has been investigated by several researchers, including Ueberreiter [11] and Krasicky et al. [12]. In the liquid or solution state, the physical and chemical interaction of the two polymers, starch and synthetic polymer, is facilitated by increasing the degrees of freedom of the atoms and the interaction of their functional groups. From the solubility parameters, this interaction or the compatibility between the two polymers and the solvents can be predicted, this parameter reflects the degree of its cohesive energy. In the literature, there are different proposals for determining these parameters, for example, using the Flory-Huggins solution theory, the Hildebrand solubility parameters, van Krevelen and the Hansen parameters [13–15]. The latter being the most used because it takes into account the forces of dispersion (δ_d), the hydrogen bonds (δ_H) and the polar forces (δ_p) of polymers and solvents, Eq. (1).

$$\delta^2 = \delta_d^{\,2} + \delta_p^{\,2} + \delta_H^{\,2} \tag{1}$$

where δ^2 is the solubility parameter.

In **Table 2**, solubility parameters of some synthetic polymers and some solvents determined from the Hansen equation are presented.

If the difference between the solubility parameters of the solvent and the polymer is small, the compatibility of the polymers with the solvent is favored because thermodynamically its entropy of mixing would increase as well as the decrease the value of the Gibbs free energy of the solution. A rule widely used to determine the polymer-solvent miscibility is.

$$\delta_2 - 1.1 < \delta_1 < \delta_2 + 1.1 \tag{2}$$

where δ_1 is the solubility parameter of the polymer and δ_2 is the solubility parameter of the solvent [18].

The solubility parameter of a polymer can be determined from experimental results considering the solubility parameter, the average of the range of solubility parameters of the solvents in which it is completely soluble [19].

The properties of the polymers in solution are determined by the structural characteristics of the solvated macromolecular chain. The three hydroxyl groups of the starch are very skilled at their hydration; however, the starch granule is not solvated by water molecules at room temperature (**Figure 2**).

Starch is a polymer, which is difficult to dissolve due to the complexity of its chemical structure and its spatial arrangement. However, it is slightly soluble in some solvents due to the physical interaction of the three hydroxyl groups found along its main chain or its side chains with the

Polymer	δ (cal/cm^3)$^{1/2}$	Solvent	δ (cal/cm^3)$^{1/2}$
Polytetrafluoroethylene	6.20	Diethyl ether	7.10
Polyisobutylene	7.70	Hexane	7.24
Natural rubber	8.15	Decane	7.74
Polybutadiene	8.38	n-Butanol	7.81
polyethylene	8.67	Cyclohexane	8.20
Poly(n-butylmethacrylate)	8.75	Toluene	8.88
Poly(ethylmethacrylate)	8.95	Benzene	9.02
Polystyrene	9.12	Chloroform	9.26
Poly(methylmethacrylate)	9.25	Acetone	9.74
Polycaprolactone	9.35	Octanol	10.23
Neoprene GN	9.38	Acetic acid	10.44
Poly(vinylacetate)	9.40	Pyridine	10.60
Poly(vinyl chloride)	9.55	n-Propanol	11.72
Poly(ethyleneterephthalate)	10.10	Dimethyl formamide	12.15
Polyacrylonitrile	12.75	Acetonitrile	12.54
Starch	13.25	Ethanol 99.9%	12.83
		Dimethyl sulfoxide	13.03
		Methanol	14.64
		Ethylene glycol	16.91
		Glycerol	17.63
		Water	23.40

Table 2. Values of the solubility parameters of Hansen [13, 14, 16, 17].

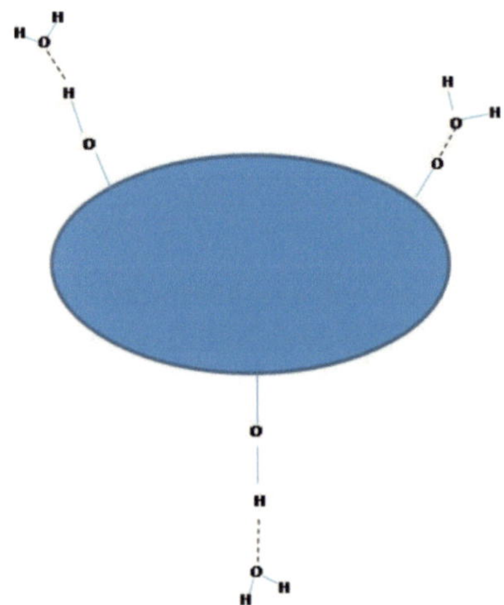

Figure 2. The starch granule remains hydrated but not solvated by water molecules at room temperature.

solvent. For example, starch is slightly soluble at room temperature in dimethyl formamide (DMF) and dimethyl sulfoxide (DMSO), partially soluble in water at temperatures above 80°C, partially in water/ethanol mixtures in the temperature range of 30–50°C and sparingly soluble in chloroform, to mention some examples. With respect to the two polymers that form the starch, the amylopectin is soluble in hot water and the amylose is not soluble (gelatinization process). Taking into account the partial solubility of starch, the search for solvents or mixtures of solvents is very important for carrying out the chemical modification of this natural polymer with synthetic polymers because in a dissolution reaction medium, it is more favored in comparison with other polymerization medium, for example, in mass and emulsion. In a dissolution reaction, there would be more chemical interaction between the two polymers, there would be no secondary reactions (the solvent would be inert), furthermore, in this reaction medium, it is easy to separate the solvent at the end of the synthesis, the released heat is absorbed by the solvent and the rate of reaction decreases, and generally, the presence of the solvent decreases the temperature of the synthesis. The investigations reported in the literature on the chemical modifications of starch with synthetic polymers are in the heterogeneous phase [20].

The solubility of the polymers is also very important to define their area of application, for example, at the biological level, they can be applied in tissue regeneration, in drug release and in membranes; at the industrial level in the manufacture of microchips and plastic materials, to mention some examples. Also, the solubility of the polymers is very important to propose its possible route of recycling.

2.3. The size of the polymer chains

Many polymers will contain molecules having many different chain lengths. It is known that the size of the chains of a polymer is one of the main factors that determines or governs its physicochemical properties, its processing behavior and consequently its possible applications.

For example, properties such as impact strength, viscosity and elasticity are in function of short chain length, medium chain length and long or large chain length, respectively [21]. The polydispersity index (PI) of a polymer is an important parameter to know the homogeneity or heterogeneity of the length of the chains of a polymer. The PI is determined by Eq. (3).

$$PI = Mn/Mw = Mz/Mw \tag{3}$$

where Mn, Mw and Mz are the number average molar mass, mass average molecular mass and mass average molecular mass-mass, respectively. A value of PI ≥1 indicates polydispersity and a value of PI = 1 means that the length of the polymer chains or molecular weight is equal. The molar masses of Mn, Mw and Mz are governed by the mechanism of the polymerization reaction and by the conditions under which it was carried out [22]. This represents a steric hindrance between the two polymers for that the chemical bond between them can be given [23, 24].

Thus, it is easier to graft or anchor on the surface of the starch a chain of smaller size of the synthetic polymer than a chain of large size due to its steric hindrance or the resulting degrees of freedom with the use of a compatibilizer (**Figure 3**).

This compatibilizer will facilitate chemical and physical interaction with larger size chains of the synthetic polymer. For example, Mani et al. [25] synthesized the starch-g-PCL compatibilizer to improve the compatibility with blends of starch or PCL; they found that their mechanical properties are higher than that of a starch/PCL blend. On the other hand, Wootthikanokkhan et al. [26] obtained the starch-g-PLA compatibilizer, which improved its compatibility with starch or PLA and the physical properties of starch.

2.4. Catalyst

The selectivity or development of a catalyst for synthesizing graft copolymers may vary greatly depending on the synthetic polymer to be used. This selectivity also depends on the temperature, pressure, concentration, particle size and chemical nature of the catalyst, whereby a copolymerization reaction is catalyzed under certain specific conditions.

A particular catalyst does not act in the same way in all the reactions on which it can act. It is of vital importance to understand the catalytic phenomenon to maintain the stability and

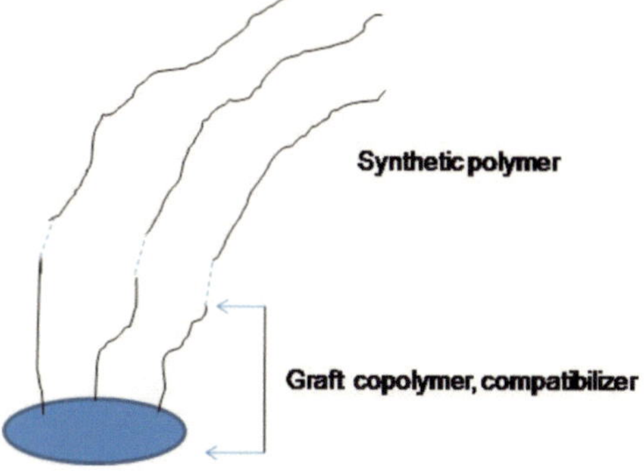

Figure 3. Compatibilizer with synthetic starch-g-polymer.

activity of the catalyst during the copolymerization as well as its process of its separation from the reactor once the copolymer is obtained. The chemical nature of the catalyst plays an important role in proposing a mechanism of copolymerization, separation or purification of the product obtained.

A variety of catalysts have been used in the synthesis of the starch-g-polymer synthetic graft copolymer. Some catalysts used in the synthesis of starch graft copolymer with synthetic polymers are presented in **Table 3**.

The graft copolymerization reactions of the starch are mainly in heterogeneous media (different phases), i.e. by heterogeneous catalysis. The choice of a catalyst for the synthesis of the graft copolymer is mainly based on the function of the synthetic polymer used. Generally, the catalyst

Graft copolymer	Catalyst	Yield (%)	Reference
Starch-g-lactic acid	Sodium hydroxide, NaOH	33.6	[27]
	Ammonia water, NH_4OH	58.9	[28]
	Sulfuric acid, H_2SO_4	75.0	[29]
Starch-g-PCL	Tin(II) 2-ethylhexanoate, $Sn(Oct)_2$	40.0	[30]
	Trimethyl aluminum, $Al_2(CH_3)_6$	58.0	[31]
	Molybdenum compounds	84.0	[32]
Starch-g-PMA	Ceric ammonium nitrate, $H_8N_8CeO_{18}$	41.0	[33]
	Ammonium nitrate, NH_4NO_3	58.0	[34]
	Enzyme horseradish peroxidase, HRP	45.0	[35]
Starch-g-PS	1-ethyl-3-methylimidazolium acetate ([EMIM]Ac)	100.0	[36]
	Potassium persulfate, $K_2S_2O_8$	66.0	[37]
	Potassium persulfate/amine	32.5	[38]
Starch-g-PAN	Ceric ammonium nitrate, $H_8N_8CeO_{18}$	94.7	[39]
	Ceric ammonium nitrate, $H_8N_8CeO_{18}$/sodium hydroxide	87.3	[40]
Starch-g-PB	Potassium persulfate, $K_2S_2O_8$	10.0	[41]
Starch-g-PGA	Sodium hydroxide, NaOH	4.2	[42]
Starch-g-PLA	Tin(II) 2-ethylhexanoate, $Sn(Oct)_2$	52.0	[43]
Starch-g-PBA	Enzyme horseradish peroxidase, HRP, hydrogen peroxide (H_2O_2) and acetyl acetone (Acac)	5.6	[44]
Starch-g-poly(ethylacrylate)	Warm distilled water	27.5	[45]
Starch-g-poly-(N-methylacrylamide-co-acrylic acid)	Potassium perdisulfate, $K_2S_2O_8$	85.4	[46]

Table 3. Some catalysts used in the synthesis of graft copolymers.

used in the synthesis of the synthetic polymer is employed in the synthesis of the graft copolymer. The starch/catalyst mass ratio is a function of the degree of substitution of the hydroxyl groups of the synthetic polymer. It is known that the greater the amount of catalyst, the faster the rate of synthesis of the copolymer but with smaller synthetic polymer chain length sizes [27, 47].

2.5. Temperature

Stability of the starch against temperature is a very important factor to consider for the synthesis of the graft copolymer. It is known that the starch granule begins to fracture or degrade at 85°C in the presence of water and temperature above this range increases the degradation. The formation of starch films by the casting method occurs at about 85°C (gelatinization temperature) in approximately 20 min. On the other hand, it is possible to carry out graft-type copolymerization at temperatures higher than 100°C with the starch granules having thermal stability at these temperatures.

This is mainly because the heat generated at these temperatures is absorbed by the other components (reactants or catalyst) of the copolymerization while the activation energy is reached to convert from reactants to products, **Figure 4**.

The starch granules are prone to initiate its degradation when the copolymerization synthesis is finalized. This is because it is very important to determine the time of synthesis to avoid the degradation of the starch granule and the generation of ashes; these ashes would generate impurities and then cause darkening of the product. The temperature and the reaction time are primarily a function of the chemical nature of the synthetic polymer and the catalyst as well as of its concentration in the reaction medium. For example, in the synthesis of the starch-g-PGA graft copolymer, with temperature higher than 180°C, potassium hydroxide and microwave irradiation were used [42].

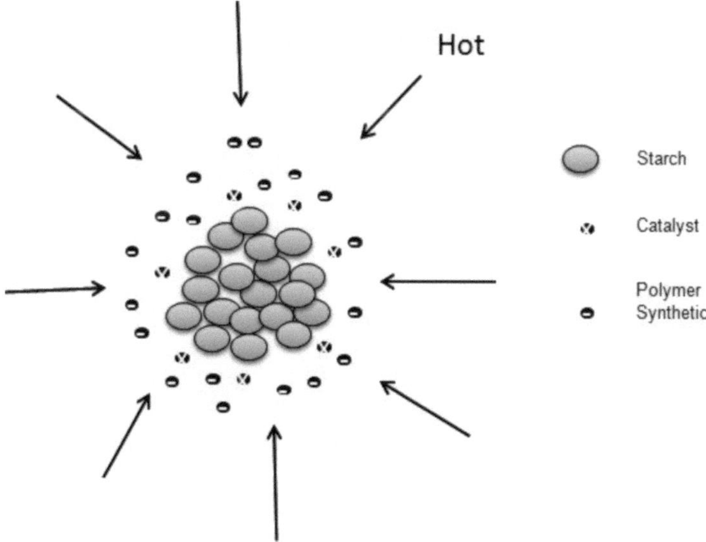

Figure 4. Thermal stability of starch.

Ramírez-Hernández et al. [32] carried out a mass polymerization for the synthesis of starch-g-PCL using a temperature range of 110–150°C. In the case of a mass copolymerization, one of the physical properties to be considered is the melting temperature of the synthetic polymer, if it presents in the reaction. This datum is important because it would allow to predict the degradation of starch granules occurs during the mass chemical reaction at the determined time. It is known that the starch begins to decompose above 85°C and if the melting temperature of the synthetic polymer is much higher, the degradation of the starch granules will be greater. Another important temperature to be known is the glass transition temperature of the synthetic polymer. If this temperature is above or below the room temperature, it would affect the physicochemical properties of the graft copolymer but this will depend on the desired mechanical properties of this copolymer.

2.6. Proportion of amylose/amylopectin and relation of the parameters of graft reactions

Starch is a polysaccharide of vegetable origin, stored in the form of granules, of complex structure found mainly in cereals (maize, rice, wheat, etc.), tubers (potatoes, sweet potatoes, yucca, etc.), legumes (pea, bean, chickpea, etc.) and fruits (mango, banana, etc.) but are also found on stems, leaves and even pollen. This semicrystalline polysaccharide consists of two polymers of α-D-glucose called amylose and amylopectin. The proportion of amylose/amylopectin and the form of the granules determines the physicochemical and functional properties of the starch granules, and consequently their use. For example, high amylose content facilitates film formation and chemical modification, but is poorly soluble in water and is more difficult to degrade with respect to a high amylopectin starch. **Table 4** shows some ratios of amylose/amylopectin reported in the literature.

According to the literature, a simple helical structure for amylose has been proposed and in the case of amylopectin, a double helix (see **Figure 5**). This arrangement of amylopectin generates an ordering of the helices, and therefore, favors its crystalline zones and decreases free space. In the case of amylose, the simple helix favors the amorphous zones with more free space between the polymer chains.

These free spaces favor the interaction with other polymer chains as the synthetic polymer. From the granule form, the amylose/amylopectin ratio and its diameter, the area of surface

Source	Moisture (%)	Starch (%)	Proteins (%)	Lipids (%)	Ashes (%)	Amylose (%)	Amylopectin (%)
Rice	16.0	83.60	0.45	0.80	0.50	25.0	75.0
Corn	13.0	85.92	0.35	0.60	0.10	27.0	63.0
Wheat	14.0	84.59	0.40	0.80	0.15	25.6	74.4
Potato	19.0	80.41	0.06	0.050	0.40	21.0	69.0
Plantain	6.6	90.00	1.60	1.70	0.10	35.0	65.0

Table 4. Amylose/amylopectin content [44–46, 48, 49].

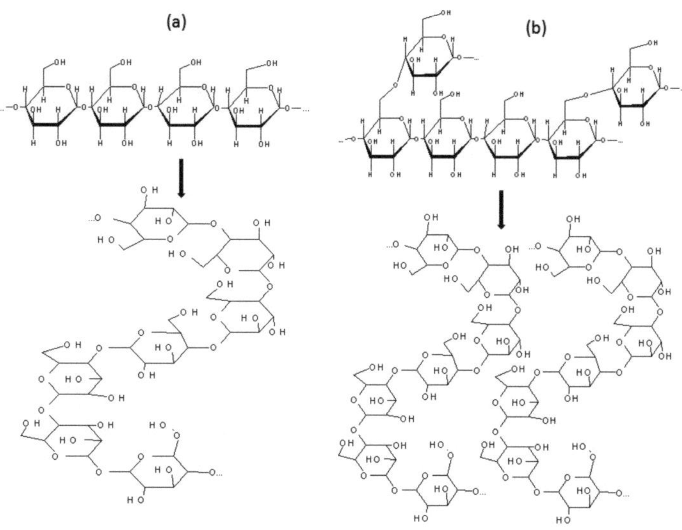

Figure 5. Structure of amylose (a) and amylopectin (b).

reactivity (R) of the starch granule can be estimated assuming that the amylose as the only component on the surface of the starch Eq. (4).

$$R = \log A^{*} Dp^{*} 3/1\mu^{2} \quad (4)$$

where A is the area of the granule and Dp is the degree of polymerization. Some values of R are presented in **Table 5** for some sources of starch.

These R values are an approximation because these values consider that all the granules have the same morphology and do not take into account the possible proportion of amylopectin present on the surface of the starch granule.

The relation of the parameters of graft reactions depends on two polymers to be studied, for example, starch-g-PE and starch-g-PCL are presented in **Figure 6**.

Source	Granule form	Diameter[a] (μm)	Amylose area (μm²)	R[*]
Potato	Oval	5–100	102–37,669	4.78–7.35
Corn	Spherical/polyhedron	2–30	50–11,309	4.47–6.83
Wheat	Spherical	2–10	50–1256	4.47–5.87
Rice	polyhedron/oval	3–12	113–1809	4.83–6.03

[*]Using a Dp = 200.
[a]Alcázar-Alay and Almeida Meireles [49].

Table 5. Values of the area of surface reactivity (R) of the starch granules.

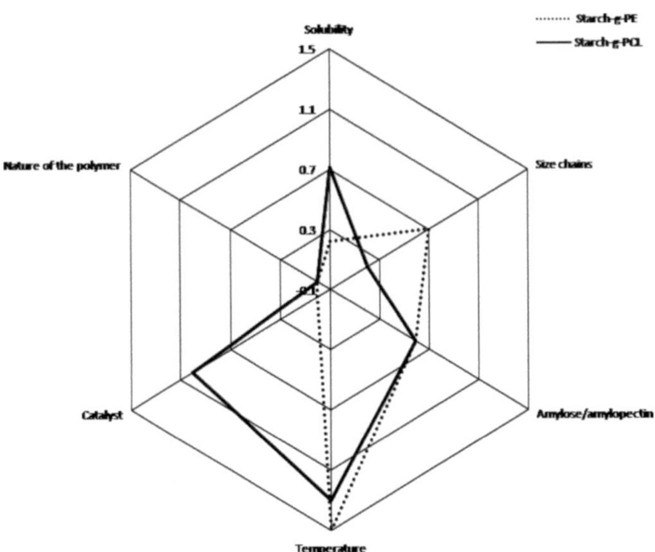

Figure 6. Relation of the parameters of graft reactions.

The solubility of the two graft copolymers depends mainly on the size of the chains of the polymer grafted on the surface of the starch granule and the chemical nature. The greater the chain length, the lesser is the solubility and vice versa, the shorter its chain length major is its solubility. This is mainly due to steric hindrance to the hydrophilic character and hydrophobic nature of the two polymer chains. The starch-g-PE graft copolymer has less solubility and its grafted chain length is greater than the starch-g-PCL. The proportion of amylose present in the starch source is important because this favors the graft reaction; in this case, both copolymers are obtained using the same source of the starch (banana starch) and the same amylose/amylopectin ratio. The presence of a catalyst in the copolymerization reaction favors the decrease of its synthesis temperature. In the case of the synthesis of the starch-g-PCL copolymer when a catalyst is used, the synthesis temperature is lower compared to the synthesis of the starch-g-PE which does not employ a catalyst and this uses a higher synthesis temperature. In both copolymers, a synthetic and a natural polymer is used; grafting a natural polymer onto the surface of the starch granule would be very difficult due to its steric hindrance since these are generally polymers of high molar mass and large chain length.

2.7. Chemical identification of the graft copolymer

In the synthesis of a graft copolymer of starch-g-synthetic polymer, the verification that the chemical reaction was carried out between the natural polymer and the synthetic polymer is of vital importance because of this experimental evidence, it is possible to explain all the physicochemical properties of the graft copolymer. There are a variety of instrumental techniques reported in the literature to characterize the chemical bond of graft type. The most common is attenuated total reflection Fourier transform infrared spectroscopy (ATR-FTIR). FTIR allows a quick and qualitative identification of the functional groups present in the copolymer; however, it is very difficult to identify from this technique, a chemical bond resulting from the union between two polymers as the starch and synthetic polymer. One of the possible reasons is the amount of this chemical bond in the graft copolymer compared with the rest of the chemical bonds of the copolymer is very small. The intensity of this signal would be expected to be weak and could be overlapped with some other vibration of the copolymer. The vibration signals of the α(1–4) and α(1–6) bonds of the starch are approximately at 1024 and 994 cm^{-1}, respectively. These two vibrations correspond to carbon and oxygen (C—O) interaction [32, 50]. In the case of graft copolymerization involving the carbon (C) atoms of the synthetic polymer and the oxygen (O) atoms of the starch, the vibration signal of this chemical bond would become overlapped. However, the decrease in the intensity of some signals of the graft copolymer spectrum, such as the hydroxyl (OH) group of the starch compared to that of the homopolymers (previously to this analysis, the separation of homopolymers, which did not react and the catalyst), this would indicate that very probably the chemical bond existence between the natural and synthetic polymer, but it is a necessary assistance of other instrumental techniques such as nuclear magnetic resonance (NMR). One of the requirements for obtaining a liquid NMR spectrum is the solubility of the sample in some deuterated solvents. In the case of starch, it is a polymer with high molecular weight, which generates a low solubility in common deuterated solvents such as deuterated chloroform and deuterated water; however, it is partially soluble in DMSO. In the literature, a variety of solvents have been reported for obtaining a magnetic resonance spectrum of starch, but in many cases, the resolution of these is poor. An NMR spectrum (^{1}H and ^{13}C) would allow to identify with certainty whether there was a chemical bond between the synthetic polymer and the natural polymer. Two of the typical signals of a ^{1}H NMR spectrum of the starch are at 5.11 and 4.57 ppm, which correspond to the α(1–4) and α(1–6) bonds, respectively. These two signals in a spectrum of ^{13}C NMR exhibit a chemical shift of 100.45 and 64 ppm, respectively. In this same spectrum, a chemical shift between 100 and 175 ppm could represent the chemical bond of the graft copolymer, in the case of an interaction of carbon (C) and oxygen (O) [51, 52]. Scanning electron spectroscopy (SEM) allows the identification of changes in the morphology of the starch granule in the case of a graft copolymer or if it is only a physical mixture of the two polymers. In the SEM micrographs reported in the literature, it has been found that the graft copolymer of starch-g-synthetic polymer favors the formation of clusters or aggregates, whereas in a blend, the synthetic polymer covers the surface of the starch granules due to its physical interaction with the synthetic polymer [20]. The X-ray diffraction technique will allow to analyze the increase or decrease in the crystallinity of the starch or the synthetic polymer as a result of its chemical interaction. In the literature, it has been reported that a decrease in the crystallinity of the synthetic polymer is an indicative of the presence of the graft copolymer [53, 54]. The instrumental technique of thermogravimetric analysis and differential scanning calorimetry would corroborate a decrease in the crystallinity of the graft copolymer with respect to that of homopolymers through the decrease in the intensity of the signals and its displacement, for example, in the decomposition and melting temperatures. The BET technique allows to verify a decrease in the surface area of the starch granule due to the synthetic polymer grafts onto it [55–57]. There are other instrumental and physicochemical techniques that allow to complement the verification of the chemical bond of the starch with the synthetic polymer through the measurement of its physical and chemical properties. The chemical bond of the starch with the synthetic polymer is necessary to corroborate it through a set of instrumental techniques and not only using one of these techniques.

2.8. Applications of the starch-g-synthetic polymer graft copolymer

Worldwide research is being carried out to generate materials more environmentally friendly and that these become an alternative to compete with conventional plastics. The synthesis of graft-type copolymers between starch and synthetic polymers is a serious alternative to obtain biodegradable materials with good physicochemical properties. The applications of these copolymers are widely depending on the source of starch and the synthetic polymer (**Table 6**).

Graft copolymers	Applies	References
Starch-g-lactic acid	Material with excellent degradability	[27–29]
Starch-g-PCL	In food packaging, drug delivery, bags and flavored biodegradable materials	[30–32]
Starch-g-PMA	Food packing, biomedical fields, coating and adhesives, drag reduction, textile industry and preparing biodegradable hydrogels and agricultural mulch films	[33–35, 58]
Starch-g-PS	in the production of paper, textiles and food additives, as well as being a commonly used biotemplate and carbon precursor in materials research	[36–38]
Starch-g-PAN	In the field of biomedicine and pharmacy such as soft contact lenses, super absorbents, drug-delivery system, polymeric dispersions, pigments and improved high gloss papers coated with the coating color compositions, soil conditioners, additives for paper and textiles, adhesives, enhanced oil recovery and sanitary goods	[39, 40, 59]
Starch-g-PB	Also disclosed are paper coating color compositions comprising the polymeric dispersions and pigments and improved high gloss papers coated with the coating color compositions	[41]
Starch-g-PGA	Copolymer allows it to have many potential applications in drug delivery, food, water treatment, cosmetics and other fields. More detailed rheological property studies of starch-PGA graft copolymers are needed to explore their potential applications	[42]
Starch-g-PLA	An expected improvement in their processability and biocompatibility due to PLA branches can be utilized in a wide range of engineering, bioengineering and medical application areas, which will be the subject of our future investigations	[43]
Starch-g-PBA	This material provides an attractive alternative for the preparation of modified starch for industrial applications how a thermal stabilizer	[44, 60]
Starch-g-poly(ethylacrylate)	The sorbent could be used successfully used for five consecutive adsorption–desorption cycles, which indicated its high reusability	[45]
Starch-g-poly-(N-methylacrylamide-co-acrylic acid)	The graft copolymer can be used as a biodegradable Hg (II) ions adsorbing agent for the removal of toxic Hg (II) ions from the waste water enriched with Hg (II) ions.	[46]
Starch-g-PVA/HA	Biomedical applications	[61]
Starch-g-AAm	Waste water treatment (flocculants), polyvalent metal cation sorbents, biodegradable superabsorbent for hygienic, agricultural and cosmetics purposes, paper industry, textile industry, petrochemical and mineral recovery	[62]
Starch-g-poly(benzyl methacrylate)	Environmentally friendly materials which are promising materials such as fillers, stabilizers, modifiers, matrices or plastics which can replace nonbiodegradable and compostable petroleum-based plastic materials	[63]

Table 6. Applications of some starch-g-synthetic polymer graft copolymers.

3. Conclusions

Synthesis of starch-g-synthetic polymer graft copolymers is increasing worldwide because this copolymer is an alternative for the generation of biodegradable materials and good physicochemical properties to compete with conventional plastics. In order to carry out this synthesis, it is necessary to review several factors such as the chemical and physical properties of the homopolymers to propose the most suitable conditions for the copolymerization reaction. The solubility of the two homopolymers in a solvent is very important to facilitate the graft copolymerization and its chemical characterization. However, the higher the solubility, the lower is the value of the molar mass grafted onto the starch granule, that is, the lengths of the grafted polymer chains decrease. The use of homopolymers of similar nature and the presence of a catalyst favor the copolymerization reaction and the decrease in their synthesis temperature, respectively. Synthesis of starch copolymers with polyhydrocarbons and hydrophobic polymers is limited; however, graft compatibilizers of starch-g-synthetic polymers can be synthesized to favor their copolymerization. The chemical characterization of the graft copolymer must be carried out through a set of instrumental techniques to corroborate the chemical bond of the starch with the synthetic polymer. The knowledge about this type of synthesis is very promising for the reduction of synthetic polymers discarded into the environment.

Acknowledgements

We are grateful to Universidad del Papaloapan and Martha Rocio Valencia Estacio for their assistance on this chapter.

Author details

Aurelio Ramírez Hernández

Address all correspondence to: chino_raha@hotmail.com

Departamento de Química, Universidad del Papaloapan, Tuxtepec, Oaxaca, México

References

[1] Tai NL, Raju-Adhikari SR, Adhikari B. Flexible starch-polyurethane films: Physiochemical characteristics and hydrophobicity. Carbohydrate Polymers. 2017;**163**:236-246

[2] Dang MN, Thi VV, Anne-Cecile G, Huy HT, Chi NHT. Biodegradability of polymer film based on low density polyethylene and cassava starch. International Biodeterioration & Biodegradation. 2016;**115**:257-265

[3] Sabetzadeh M, Bagheri R, Masoomi M. Effect of nanoclay on the properties of low density polyethylene/linear low density polyethylene/thermoplastic starch blend films. Carbohydrate Polymers. 2016;**141**:75-81

[4] Sabetzadeh M, Bagheri R, Masoomi M. Study on ternary low density polyethylene/linear low density polyethylene/thermoplastic starch blend films. Carbohydrate Polymers. 2015; **119**:126-133

[5] Bikiaris D, Prinos J, Koutsopoulos K, Vouroutzis N, Pavlidou E, Frangis N, Panayiotou C. LDPE/plasticized starch blends containing PE-g-MA copolymer as compatibilizer. Polymer Degradation and Stability. 1998;**59**(1):287-291

[6] Kittisak J, Noppol L, Phisit S, Somchai W, Charin T, Toshiaki O. Reactive blending of thermoplastic starch and polyethylene-graft-maleic anhydride with chitosan as compatibilizer. Carbohydrate Polymers. 2016;**153**:89-95

[7] Ritter W, Gardenier KJ, Kempf W. German Offen. DE 4 038 700; 1992.

[8] Pereira CS, Cuncha AM, Reise RL, Vázquez B, Sanroman J. New starch-based thermoplastic hydrogels for use as bone cements or drug-delivery carriers. Journal of Materials Science Materials in Medicine. 1998;**9**(12):825-833

[9] Miller-Chou BA, Koenig JL. A review of polymer dissolution. Progress in Polymer Science. 2003;**28**(8):1223-1270

[10] Billmeyer FW. Ciencia de los polímeros. Editorial Reverte, 23. España; 2004

[11] Ueberreiter K. The solution process. In: Crank J, Park GS, editors. Diffusion in Polymers. New York, NY: Academic Press; 1968. pp. 219-257

[12] Krasicky PD, Groele RJ, Rodriguez F. Measuring and modeling the transition layer during the dissolution of glassy polymer films. Journal of Applied Polymer Science. 1988;**35**(3):641-651

[13] Hamaide T, Deterre R, Feller JF. Environmental Impact of Polymers. 10. United kingdom: Editorial Wiley; 2014

[14] Walter RH. Polysaccharide Association Structures in Food. 100. USA; 1998

[15] Barton A. Handbook of Solubility, Parameters and Other Cohesion Parameters. 2nd ed. New York: CRC Press; 1991. pp. 157-193

[16] Small PA. Some factors affecting the solubility of polymers. Journal of Applied Chemistry. 1953;**3**:71-80

[17] Cecopieri-Gómez ML, Palacios J. Cálculo teórico y experimental del parámetro termodinámico de interacción de Flory del Poli(adipato de etileno). Revista de la Sociedad Química de México. 2001;**45**(2):82-88

[18] Sun SF. Physicalchemistry of Macromolecules: Basicprinciples and Issues. USA: Wiley; 2004

[19] Laura F, Villamizar LF, Martínez F. Physicochemical study of the solubility of eudragit s100® in some aqueous and organic systems. Revista Colombiana de Quimica. 2008;**37** (2):173-187

[20] Ramírez-Hernández A, Mata-Mata JL, Aparicio-Saguilán A, González-García G, Hernández-Mendoza H, Gutiérrez-Fuentes A, Báez-García JE. The effect of ethylene glycol on starch-g-PCL graft copolymer synthesis. Starch-Starke. 2016;**68**:1148-1157

[21] Krevelen DW, Nijenhuis K. Properties of Polymers. Amsterdam: Elservier; 2009

[22] Rogošić M, Mencer H, Gomzi Z. Polydispersity index and molecular weight distributions of polymers. European Polymer Journal. 1996;**32**(11):1337-1344

[23] Mua JP, Jackson DS. Fine structure of corn amylase and amylopectin fractions with various molecular weights. Journal of Agricultural and Food Chemistry. 1997;**45**:3840-3847

[24] Yoo SH, Jane JL. Molecular weights and gyration radii of amylopectins determined by high-performance size-exclusion chromatography equipped with multiangle laser-light scattering and refractive index detectors. Carbohydrate Polymers. 2002;**49**:307-314

[25] Mani R, Tang J, Bhattacharya M. Synthesis and characterization of starch-graft-polycaprolactone as compatibilizer for starch/polycarprolactone blends. Macromolecular Rapid Communications. 1998;**19**(6):283-286

[26] Wootthikanokkhan J, Kasemwananimit P, Sombatsompop N, Kositchaiyong A, Isarankurana S, Kaabbuathong N. Preparation of modified starch-grafted poly(lactic acid) and a study on compatibilizing efficacy of the copolymers in poly(lactic acid)/thermoplastic starch blends. Journal of Applied Polymer Science. 2012;**126**:389-396

[27] Qingling W, Yingmo H, Jianhua Z, Yang L, Xue Y, Jing B. Convenient synthetic method of starch/lactic acid graft copolymer catalyzed with sodium hydroxide. Bulletin of Materials Science. 2012;**35**(3):415-418

[28] Yingmo H, Mingru T. Synthesis of starch-g-lactic acid copolymer with high grafting degree catalyzed by ammonia water. Carbohydrate Polymers. 2015;**118**:79-82

[29] Hu Z. The optimized synthesis of starch-g-lactic acid copolymer with high grafting degree catalyzed by sulfuric acid. Journal of Wuhan University of Technology—Materials Science Edition. 2014;**29**(6):1187-1190

[30] Chen L, Nia Y, Biana X, Xueyu Q, Zhuanga X, Chena X, Jinga X. A novel approach to grafting polymerization of ε-caprolactone onto starch granules. Carbohydrate Polymers. 2005;**60**(1):103-109

[31] Dubois P, Krishnan M, Ramani N. Aliphatic polyester-grafted starch-like polysaccharides by ring-opening polymerization. Polymer. 1999;**40**:3091-3100

[32] Ramírez-Hernández A, Aparicio-Saguilán A, Mata-Mata JL, González-García G, Hernández-Mendoza H, Gutiérrez-Fuentes A, Báez-García E. Chemical modification of banana starch by the in situ polymerization of e-caprolactone in one step. Starch-Starke. 2016;**69**(5). DOI: 10.1002/star.201600197

[33] De Bruyn H, Sprong E, Gaborieau M, David G, Roper JA, Gilbert RG. Starch-graft-copolymer latexes initiated and stabilized by ozonolyzed amylopectin. Journal of Polymer Science Part A: Polymer Chemistry. 2006;**44**(20):5832-5845

[34] Bemiller J, Whistler R. Starch: Chemistry and Technology. Vol. 729. 2009

[35] Wang S, Wang Q, Xuerong F, Jin X, Ying Z, Jiugang Y, Heling J, Cavaco-Paulo A. Synthesis and characterization of starch-poly(methyl acrylate) graft copolymers using horseradish peroxidase. Carbohydrate Polymers. 2016;**136**:1010-1016

[36] Yongjun M, Xiyan D, Jianan S, Leli W, Zhengping L. Preparation of corn starch-g-polystyrene copolymer in ionic liquid: 1-Ethyl-3-methylimidazolium acetate. Carbohydrate Polymers. 2015;**121**:348-354

[37] Chang GC, Kiho L. Preparation of starch-g-polystyrene copolymer by emulsion polymerization. Carbohydrate Polymers. 2002;**48**(2):125-130

[38] Vladimir N, Sava V, Dušan A, Aleksandar P. Biodegradation of starch–graft–polystyrene and starch–graft—Poly(methacrylic acid) copolymers in model river water. Journal of the Serbian Chemical Society. 2013;**78**(9):1425-1441

[39] Hashem A, Afifi MA, EI-Alfy EA, Hebeish A. Synthesis, characterizations and saponification of poly(AN)-starch composites and properties of their hydrogels. American Journal of Applied Sciences. 2005;**2**(3):614-621

[40] Hashem A, Afifi MA, EI-Alfy EA, Hebeish A. Synthesis, characterizations and saponification of poly(AN)-starch composites and properties of their hydrogels. American Journal of Applied Sciences. 2005;**2**(3):614-621

[41] Nguyen CC, Verne JM, Pauley EP. Patent US5130394A; 1992

[42] Jingyuan X, Krietemeyer E, Finkenstadt V, Solaiman D, Ashby R, Garcia R. Preparation of starch–poly–glutamic acid graft copolymers by microwave irradiation and the characterization of their properties. Carbohydrate Polymers. 2016;**140**:233-237

[43] Kouroush S, Mehmet Y, Zakir MOR, Erhan P. Controlled graft copolymerization of lactic acid onto starch in a supercritical carbon dioxide medium. Carbohydrate Polymers. 2014;**114**:149-156

[44] Wang S, Xu J, Wang Q, Fan X, Yu Y, Wang P, Zhang Y, Yuan J, Cavaco-Paulo A. Preparation and rheological properties of starch-g-poly(butylacrylate) catalyzed by horseradish peroxidase. Process Biochemistry. 2017;**59**:104-110

[45] Vandana S, Sadhana M. Microwave synthesis, characterization, and zinc uptake studies of starch-graft-poly(ethylacrylate). International Journal of Biological Macromolecules. 2010;**47**(3):348-355

[46] Haradhan K, Subhadip D, Tridib T. Synthesis of starch-g-poly-(N-methylacrylamide-co-acrylic acid) and its application for the removal of mercury (II) from aqueous solution by adsorption. European Polymer Journal. 2014;**58**:1-10

[47] Swanson LC, Fanta GF, Fecht RG, Burr R. Polymer applications of renewable-resource materials: Starch-g-poly(methyl acrylate) effects of graft level and molecular weight on tensile strength. Polymer Science and Technology. 1981;**17**:59-71

[48] Carballo SLM. Introducción a la catálisis heterogénea. Colombia; 2002

[49] Alcázar-Alay SC, Almeida Meireles MA. Physicochemical properties, modifications and applications of starches from different botanical sources. Food Science and Technology (Campinas). 2015;**35**(2):215-236

[50] Najemi L, Jeanmaire T, Zerroukhi A, Raihane M. Organic catalyst for ring opening polymerization of ε-caprolactone in bulk. Route to starch-graft-polycaprolactone. Starch-Starke. 2010;**62**:147-154

[51] Graaf RA, Lammers G, Janssen LPBM, Beenackers AACM. Quantitative analysis of chemically modified starches by1H-NMR spectroscopy. Starch/Stärke. 1995;**47**:469-475

[52] Cheng HN, Neiss T. Solution NMR spectroscopy of food polysaccharides. Polymer Reviews. 2012;**52**:81-114

[53] Ming L, Torsten W, Fengwei X, Frederick J, Halley P, Gilbert G. Biodegradation of starch films: The roles of molecular and crystalline structure. Carbohydrate Polymers. 2015; **122**:115-122

[54] Ming L, Torsten W, Fengwei X, Frederick J, Halley P, Gilbert G. Shear degradation of molecular, crystalline, and granular structures of starch during extrusion. Starch-Starke. 2014;**66**:595-605

[55] Vega GA, Lara AE, Lemus MR. Isotermas de adsorción en harina de maíz (*Zea mays* L.). Food Science and Tecnhology (Campinas). 2006;**26**(4):821-827

[56] Czepirski L, Komorowska-Czepirska E, Szymonska J. Fitting of different models for water vapour sorption on potato starch granules. Applied Surface Science. 2002;**196**:150-153

[57] Staudt PB, Kechinski CP, Tessaro IC, Marczak LDF, Soares R, Cardozo NSM. A new method for predicting sorption isotherms at different temperatures using the BET model. Journal of Food Engineering. 2013;**114**(1):139-145

[58] Deepak P, Reena S. Synthesis and characterization of graft copolymers of methacrylic acid onto gelatinized potato starch using chromic acid initiator in presence of air. Advanced Materials Letters. 2012;**3**(2):136-142

[59] Apopei DF, Dinu MV, Drăgan ES. Graft copolymerization of acrylonitrile onto potatoes starch by ceric ion. Digest Journal of Nanomaterials and Biostructures. 2012;**7**(2):707-716

[60] Martinez-Arellano AC, Rivera-Armenta JL, Mendoza-Martínez AM, Díaz-Zavala NP, Sandoval-Robles J, Banda-Cruz E. Study of graft copolymerization of butyl acrylate on starch using redox initiator system. Química Nova. 2014;**37**(3):426-430

[61] Lu DR, Xiao CM, Xu SJ. Starch-based completely biodegradable polymer materials. Express Polymer Letters. 2009;**3**(6):366-375

[62] Kugler S, Spychaj T, Wilpiszewska K, Gorący K, LendzionBieluń Z. Starchgraft copolymers of N-vinylformamide and acrylamide modified with montmorillonite manufactured by reactive extrusion. Journal of Applied Polymer Science. 2013;**127**(4):2847-2854

[63] Worzakowska M. Starch-g-poly(benzyl methacrylate) copolymers. Journal of Thermal Analysis and Calorimetry. 2016;**124**:1309-1318

Chapter 3

Evaluation of Styrene Content over Physical and Chemical Properties of Elastomer/TPS-EVOH/Chicken Feather Composites

María Leonor Méndez-Hernández,
José Luis Rivera-Armenta,
Zahida Sandoval-Arellano,
Beatriz Adriana Salazar-Cruz and
María Yolanda Chavez-Cinco

Additional information is available at the end of the chapter

http://dx.doi.org/10.5772/intechopen.72969

Abstract

A series of styrene-butadiene (SB) elastomer/thermoplastic starch (TPS)/ethylene vinyl alcohol copolymer (EVOH) composites were modified including chicken feathers in its formulation, which have the main component keratin. The composites were prepared by means of melt blending, and their chemical interactions were evaluated by means of infrared spectroscopy (FTIR), and their thermal properties as Tg values were investigated using differential scanning calorimetry (DSC), thermal stability using thermogravimetric analysis (TGA), and viscoelastic properties with dynamic mechanical analysis (DMA). The styrene content in SB was changed in 3 levels, and chicken feather content also changed in 3 levels. It was identified that Tg value in composites decreases that is attributed to the styrene content in elastomer and that the chicken feather improved the storage modulus of composite. The thermal stability of composites also was affected by the presence of chicken feathers due its good thermal properties.

Keywords: styrene-butadiene copolymer (SBR), EVOH, thermoplastic starch (TPS), chicken feather (CF), thermal behavior

1. Introduction

The poor disposition of solid wastes from natural resources that adversely affect the environment has developed interest related to the development of biodegradable polymers to obtain

improved composites, not only for environmental reasons but also for their properties and sustainability [1–3]. A valuable alternative with enormous potential is to manufacture composite materials using chicken feather (CF). Keratin is the main component of CF that has extraordinary properties of thermal and mechanical resistance, is also light and confers acoustic insulation, and is a light material with high mechanical and thermal resistance, and for some decades, it was the focus of attention to be used as a raw material in combination with different plastics products [4–12]. The CF that generates the poultry industry as solid waste represents more than five million tons per year. Recently, the literature points the keratin of CF as a material with low cost by its abundance in the nature and that can take advantage of the waste of the birds, which confers a status of material recycled environment friendly [10]. According to previous works, keratin can be used in conjunction with other biomaterials in medical applications suggesting affinity with cells and tissues, also in combination with other biopolymers as chitosan [13]. In addition, chicken feather keratin is a renewable and low-cost material [14–20] and is compatible with biodegradable polymers as chitosan-starch [21], cellulose, wool [4], including thermoplastic starch (TPS) [14–20] and ethylene vinyl alcohol copolymer (EVOH), which provides greater ductility and improves the elasticity of TPS. In nature, natural fibers are obtained with good physicochemical properties such as keratin, which is extracted from nature and can also be found in nails, wool, claws, horns, and feathers [8, 22]. Research has focused in conjunction with manufacturers in finding new directions for the application of keratin in polymer blends which have not been well known to the present. According to Cheng et al. [23], the feasibility of using the fiber of chicken feathers in compounds with polylactic acid (PLA) has been studied in terms of its mechanical and thermal properties. To test the mechanical and physical properties of the panels, Winandy et al. [8] have materialized a series of medium-density fiber panels containing several different blends of wood fiber and chicken feather fiber (CFF). **Figure 1** shows the constituents of a chicken feather.

Barone and Schmidt [24] reported the use of feather keratin fiber as a short fiber reinforcement in low-density polyethylene composites. Research has been reported on high-density polyethylene-based compounds using keratin from chicken feathers. Taking into account the hydrophilic properties of keratin, its potential application in the manufacture of fibers with improved sorption characteristics is seen, being useful for the production of textile material that focuses on sanitary and medical applications.

In previous works [3], chicken feather keratin fibers were used as reinforcement in the poly(methylmethacrylate) (PMMA) matrix. The composites were evaluated by thermal and dynamic-mechanical analysis. The high-temperature stability and thermal transition of the keratin/PMMA base compound were found to be higher than that of virgin PMMA.

Villarreal-Lucio et al. [25] prepared composites from chicken feathers and polyvinyl chloride (PVC) finding that composites CF-PVC represent an opportunity to obtain materials with improved properties such as thermal stability and dynamic-mechanic properties, but with a low interfacial addition between PVC and CF. Starch is a biopolymer obtained from renewable plants such potato, maize, wheat, and others sources. The main components are amylose and amylopectin, the first one is a linear polysaccharide and second one has a branched structure.

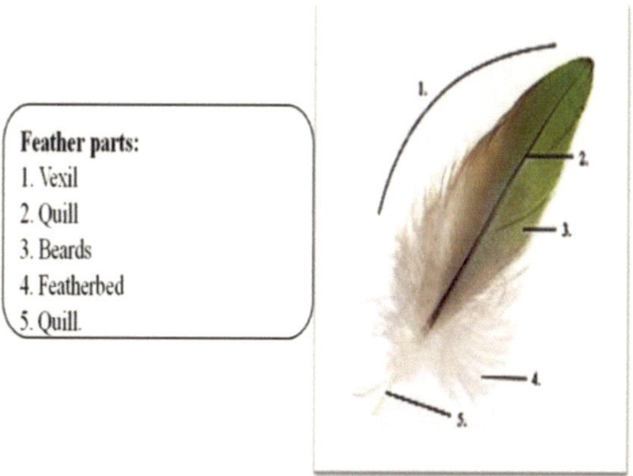

Figure 1. Constituents of a chicken feather.

It is a cheap material compared with synthetic polymers in packing applications. The TPS refers to granule starch, which has been destructurized, forming a mixture of its polymer constituents and various proteins, lipids, and smaller molecules that are also contained in the starch granule. With the aim to improve its processability and mechanical properties and moisture resistance, starch is blended by extrusion with other polymers, using plasticizers as glycerin and water. In this way, it is possible to obtain a material with desired physical properties [26]. The transformation of native starch to thermoplastic starch (TPS) by extrusion results in the loss of the natural chain structure. That is, processing in the presence of heat and water causes the starch granules to gel or break, progressively destroying crystallinity [26]. Starch has been used as a thermoplastic material because it is a renewable, biodegradable, and very low-cost resource [21]; the most effective plasticizer for starch-based materials is glycerol, water, sorbitol, and urea [26]. To improve the properties of thermoplastic starch (TPS) because it is not a good choice because of its poor mechanical properties and because of its susceptibility to moisture, the structure of the starch has been modified, mixed with other polymers (biodegradable and/or synthetic), and a better interfacial adhesion is obtained with compatibilizers. **Figure 2** shows the process of gelatinization and plasticization for starch [27].

One of the thermoplastic materials compatible with TPS is EVOH, which is widely used in the food packaging industry. It is a biodegradable polymer and considered a good compatibilizer that when added to a mixture of immiscible materials during the extrusion modifies the interfacial properties and stabilizes the mixture. It is known that EVOH provides it a better ductility and improves the elasticity of TPS [26]. Several researches have been carried out with TPS-EVOH composites reinforced with natural biopolymers, and among others are coir, cellulose fibers [15, 16], and hydroxyapatite [17–19], but there are no reports where CF is used as a reinforcer in TPS/EVOH composites. Several studies have been carried out on the composite materials using SB elastomers and keratin from the CF, which has allowed knowing about the influence of the different variables on the thermal and mechanical behavior, as well as the various valid tools for the observation of variations [2, 9, 20, 28–30]. With respect to

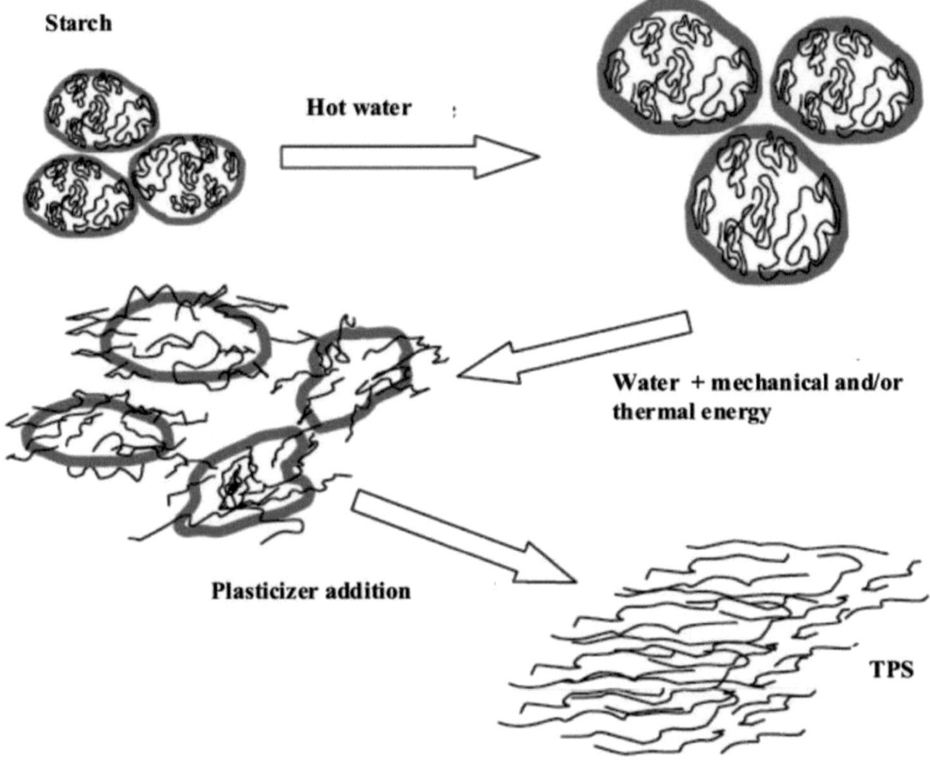

Figure 2. Gelatinization and plasticization of starch.

the above, several important aspects are listed namely dynamic-mechanical behavior, hydrophilic, acoustic, and thermal stability properties of keratin, and its study per se represents a vast field of scientific exploration. In the other hand, the use of elastomers in combination with other materials as compound has been reported, trying to improve ductility and oxygen barrier of elastomers [31–33]. Among the elastomers, the styrene-butadiene (SB) copolymers are the most valuables due to the wide application areas as adhesives, asphalt modifiers, sealants, shoe soles, impact modifiers, and also can be compounded to produce materials that enhance grip, feel, and appearance in applications such as tires, automotive parts, and packing [34, 35]. SB copolymers are considered a thermoplastic material, whose elastic behavior[1] and thermoplastic behavior[2] are combined at same time. The combination of good mechanical properties and processability makes the SB copolymers an interesting kind of materials. It is essential that hard (styrene block) and soft (butadiene) segments are immiscible on a microscopic scale. One important variable is SB copolymer properties, the styrene content, which gives a plastic characteristic to copolymer with high styrene content, and elastomer behaves as a vulcanized elastomer at low styrene content. Because of its good processability behavior, thermal, and mechanical properties, SB copolymers have been widely used in composite materials, trying

[1]Property to change and recover the shape when a force has applied and then removed.

[2]The property to become viscous and free-flowing liquid when heated and resolidified when cooled to room temperature.

Material	Elastomer (g)	TPS-EVOH (g)	Chicken feathers (g)
SB1-45/TPS-EVOH/CF	45	12	3
SB1-50/TPS-EVOH/CF	50	8	2
SB1-55/TPS-EVOH/CF	55	4	1
SB2-45/TPS-EVOH/CF	45	12	3
SB2-50/TPS-EVOH/CF	50	8	2
SB2-55/TPS-EVOH/CF	55	4	1
SB3-45/TPS-EVOH/CF	45	12	3
SB3-50/TPS-EVOH/CF	50	8	2
SB3-55/TPS-EVOH/CF	55	4	1

SB1-elastomer with 45% styrene content, SB2-elastomer with 32% styrene content, SB3-elastomer with 25% styrene content, TPS-EVOH with 75% TPS content.

Table 1. Identification codes for prepared composites.

to take advantage of its properties to obtain materials with a higher performance than conventional ones. In this work, three SB elastomers with different styrene contents were combined with a TPS-EVOH blend, and additional to that, CF was added to obtain a composite that presents enhanced properties. The effect of styrene content in SB elastomer was studied. **Table 1** shows the identification codes for prepared composites. By means of infrared spectroscopy (FTIR), the possible chemical reactions between components in composite were evaluated, and thermal properties were evaluated by means differential scanning calorimetry (DSC), dynamic mechanical analysis (DMA), and thermogravimetric analysis (TGA).

2. Experimental

2.1. Materials

Chicken feather (CF) was obtained from a local slaughterhouse in Altamira city, México, and three types of styrene-butadiene (SB) used were SB1 45% styrene content, SB2 32% styrene content, and SB3 25% styrene content, provided by Dynasol Elastomers S.A. de C.V. Chicken feathers (CFs) were cleaned with several washes, first with distilled water and then with acetone and finally with ethanol; after that CFs were dried at room temperature to be clean, sanitized, and odor free, and then proceeded to remove the barbules from quill, that was cut it into small pieces [1, 25]. A twin screw extruder ZSK30 with nine heating zones was used to obtain the mixtures of TPS-EVOH. The EVOH melted at 200°C was fed in zone 5 using the single-screw extruder Killion KTS-100 at full speed. The mixing of EVOH with TPS started from zone 5 to zone 8. In the pumping zone 9, the pressure of the extrudate was increased. The screw speed for all blends was 150 rpm. Mixtures were prepared with 75% TPS proportion [36]. **Figure 3** shows the configuration of double screw extruder to obtain the blends with EVOH.

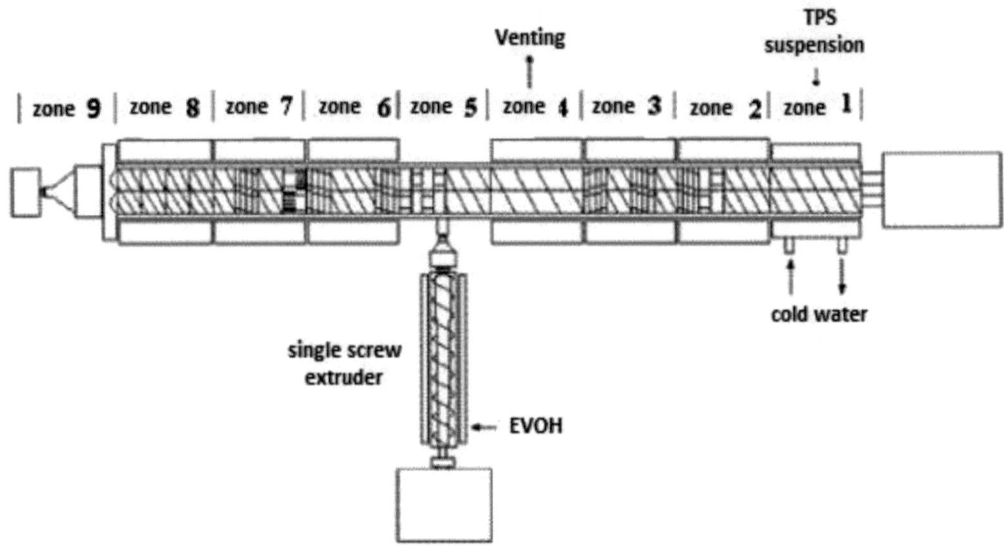

Figure 3. Configuration of double screw extruder with nine heating zones, to obtain TPS-EVOH mixtures using a single screw extruder to melt at 200°C the EVOH.

2.2. Preparation elastomer/TPS-EVOH/keratin composite

The elastomer/TPS-EVOH/keratin composites were prepared by melting the mix using a plasticorder/Brabender PL2000 torque rheometer, establishing the optimum conditions at 185°C with 20 min of mixing, using roller blades at 100 rpm speed. Then, the materials were compressed in a Dake press with 10 tons during 20 min, using appropriate molds. **Table 1** shows the codes used for identification of composites.

2.3. Composites characterization

2.3.1. *Infrared spectroscopy (FTIR)*

The infrared spectroscopy technique was used to identify functional groups of materials, and for that purpose, equipment Perkin Elmer Spectrum One model was used, with attenuated total reflectance (ATR) technique with ZnSe plates in a range of 4000–600 cm^{-1} with 12 scans.

2.3.2. *Differential scanning calorimetry (DSC)*

The differential scanning calorimetry (DSC) was used to determine the thermal transitions of the composites, using Perkin Elmer DSC8000 equipment. The employed method first consists of heating cycle from 30 to 200°C at 10°C/min, then a cooling cycle from 200°C up to −100°C, and the sample was kept for 5 min at this temperature and a second heating ramp from −100 to 200°C was carried out, with heating rate of 10°C/min, taking the second heating for analysis. The sample amount used was 10 ± 2 mg, in an inert atmosphere of nitrogen, with a flow rate of 20 mL/min.

2.3.3. *Dynamic mechanical analysis (DMA)*

The dynamic mechanical analysis was carried out in a DMA-Q800 TA-Instrument, with a double cantilever clamp, and sample dimensions were 30 × 12 × 3 mm in a rectangular shape (length, wide, thickness). Analysis were carried out in multifrequency mode with temperature range from −100 to 200°C, with a heating rate 5°C min^{-1}, 1 Hz frequency, and amplitude 2 μm.

2.3.4. *Thermogravimetric analysis (TGA)*

The thermogravimetric analysis was carried out in an SDT (DSC-TGA) Q600 TA Instrument equipment under an N_2 atmosphere with 100 mL/min flow, in a temperature range from 40 to 600°C with 10°C/min rate, using 10 ± 2 mg sample.

3. Results

3.1. Infrared spectroscopy of elastomer/TPS-EVOH/keratin composite

Figure 4 shows the IR spectra for SB3, TPS-EVOH, CF, and SB3/TPS-EVOH/CF composites from 4000 to 600 cm^{-1}. It is possible to identify some of the functional groups in the blends of the elastomers with TPS-EVOH and CF. The SB3 signals at 3000 and 3100 cm^{-1} are associated with unsaturated carbons; meanwhile, at 2900 and 2850 cm^{-1}, the signals are related to the

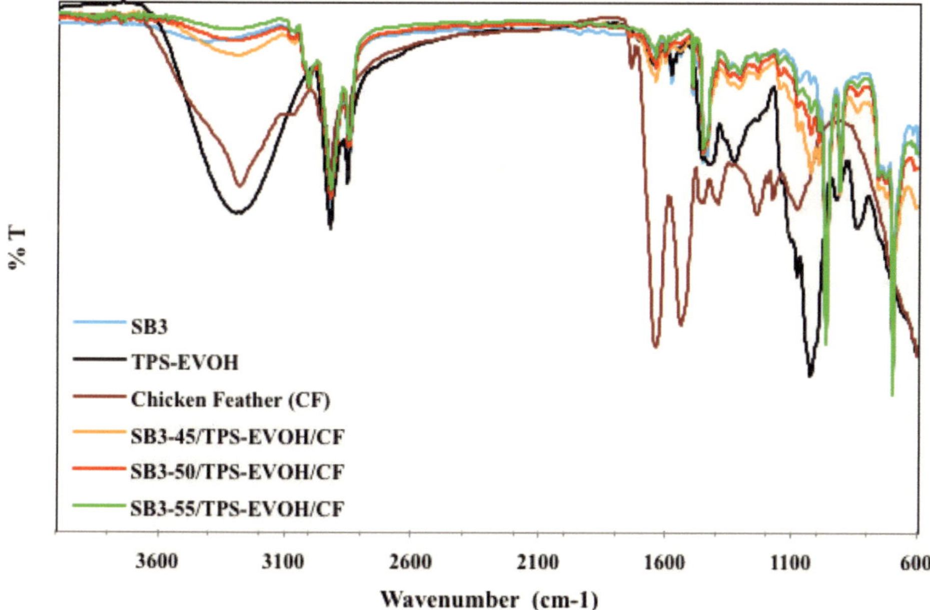

Figure 4. FTIR spectra of SB3, TPS-EVOH, CF, and SB3/TPS-EVOH/CF composite.

stretching of methyl and methylene groups; besides, the region of the aromatic ring is visible from 2000 to 1850 cm^{-1} [35]. The CF signals at 3300 cm^{-1} correspond to the ordered regions of NH group of amides A α-helix conformation, at 2950 cm^{-1} is related to the asymmetry vibration of CH group of methyl, and the band at 1710 cm^{-1} is matched to the vibration of amides I of β-sheet conformation, for C=O group of amides I α-helix at 1650 cm^{-1} and at 700 cm^{-1} attributed to the vibration of C—S group. The main groups assigned at 1650 and 1550 cm^{-1} peaks from the chicken feather keratin are amide I and amide II bands, respectively. The peaks at 1500, 1450, and 1250 cm^{-1} are attributed to a plane bending of NH group that corresponds to β-sheet conformation, the bending of $—CH_3$ group, and CN group of amides III, respectively [37]. The glycerol absorption peaks are at 1109, 1042 and 994 cm^{-1}. The signals at 3330 cm^{-1} correspond to hydroxyl stretching of EVOH and are evident by the high amount of hydroxyl groups in the chemical structure [38]. The vibrations at 1150, 1100 and 1050 cm^{-1} are assigned to vibrations of C—C group, a peak around 700 cm^{-1} is attributed to vibration of C—S group, and finally, vibrations at 970, 910, 760, and 690 cm^{-1} show the evidence of unsaturated aromatic carbon deformations [24, 39, 40]. Composites SB/TPS-EVOH/CF show typically the same signals, and the variations are attributed to elastomer content, and it is not possible to identify any evidence of a chemical reaction between materials.

3.2. Differential scanning calorimetry (DSC)

Table 2 presents DSC results for elastomers, CF and for composites. It was observed that CF showed two transitions, one around 140°C and the second one around 260°C, attributed to the crystalline melting temperature of keratin, the main component of CF, corresponding with

Material	Transition (°C)		
Chicken feather (CF)		140	263
SB1	−40		
SB1-45/TPS-EVOH/CF	−82	58	162
SB1-50/TPS-EVOH/CF	−85	48	169
SB1-55/TPS-EVOH/CF	−92	6	175
SB2	−33		
SB2-45/TPS-EVOH/CF	−88	48	158
SB2-50/TPS-EVOH/CF	−89	39	164
SB2-55/TPS-EVOH/CF	−95	31	182
SB3	−63		
SB3-45/TPS-EVOH/CF	−83	131	165
SB3-50/TPS-EVOH/CF	−85	28	159
SB3-55/TPS-EVOH/CF	−90	−4	151

Table 2. Transition value for SB/TPS-EVOH/CF composites obtained by DSC.

previous reports [8]. Other reports indicate that CF did not show any melting peak [18], and also there are reports that around 300°C the disulfide bonds and denaturation of helix structure of keratin occurs [21]. Other reports [9] are known that Tg value of keratin is affected by water concentration and the content of alpha and beta keratin. The SB elastomers only show a Tg value which depends on the styrene content varying from −40°C for elastomer with higher styrene content, and −63°C for elastomer with lower styrene content. These results are according with soft and hard segments of copolymer [41]. The SB-TPS/EVOH/CF composites show two Tg values, one at low temperatures (around −90°C), the second one depends on styrene content in elastomer, 48°C for higher styrene content in elastomer and 28°C for lower styrene content, which is indicative that there is an effect of hard segments of styrene in composite structure, due to the concentration of keratin was constant. Another interesting point is that lower Tg value (attributed to butadiene segments in elastomer) also changed in SB/TPS-EVOH/CF composite moving at a temperature around −94 to −85°C, indicative that soft segments in composites are affected. These results indicate that the softness of materials is improved and the molecular level, produced by polypeptide chains of keratin with elastomer structure. A third transition was identified in SB/TPS-EVOH/CF composites around 160°C, which is not been reported before, however as was discussed before, can be attributed to keratin and the diminish is due the water content, in this case OH groups from TPS and EVOH which effects on keratin structure providing to material a soft characteristic.

Janowska et al. [9] reports that DSC technique cannot be a good option to evaluate Tg due to their considerable thermal effect caused by evaporation of physically combined materials, in this case TPS and EVOH, which can mask the results due to a considerable effect of water content. Nevertheless, it can be possible to identify an effect of styrene content in an elastomer over Tg of composites.

3.3. Dynamic mechanical analysis (DMA)

DMA is a useful technique to determine the viscoelastic properties of composite materials related to primary relaxations and other parameters. Dynamic mechanical properties are used for detection of damping peaks, elastic, and loss modulus changes with temperature [26]. Dynamic mechanical analysis (DMA) for TPS and blends is a useful tool to evaluate the phase separation relaxations and viscoelastic properties [42]. DMA can find a low-temperature peak around −40°C attributed to the secondary relaxation of starch and corresponds to the secondary transition of glycerol-rich domains, whereas a higher temperature peak is due to the primary transition of amylopectin-rich domain [42]. DMA was performed to evaluate the effect of elastomer type on an SB/TPS-EVOH/CF composite. **Figure 5** shows the storage modulus (E′) versus temperature for SB1, TPS-EVOH, and SB1/TPS-EVOH/CF composites. The dynamic mechanical properties of TPS blends have been studied, TPS can form immiscible blends due to the high interfacial tension of polar and nonpolar segments of components. TPS-EVOH shows two relaxations around −40°C and 35°C, attributed to both components TPS and EVOH, which are partially miscible (see tan δ curve for TPS in **Figure 6**). High concentrations of plasticizer cause a heterogeneous system, resulting from glycerol and starch domains in compounds [43, 44]. According to results, for the composite SB1-45/TPS-EVOH/CF (**Figure 5**) the storage modulus has a good thermomechanical behavior compared to the compound with the highest content of SB (SB1-55/TPS-EVOH/CF), where the pure SB1 elastomer has a acceptable thermomechanical

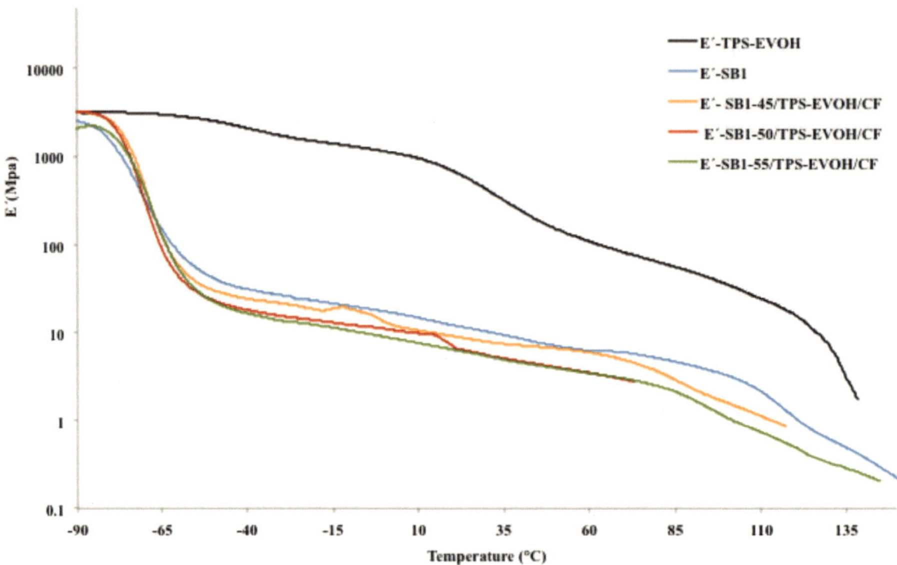

Figure 5. Storage modulus curve from DMA for SB1/TPS-EVOH/CF composites.

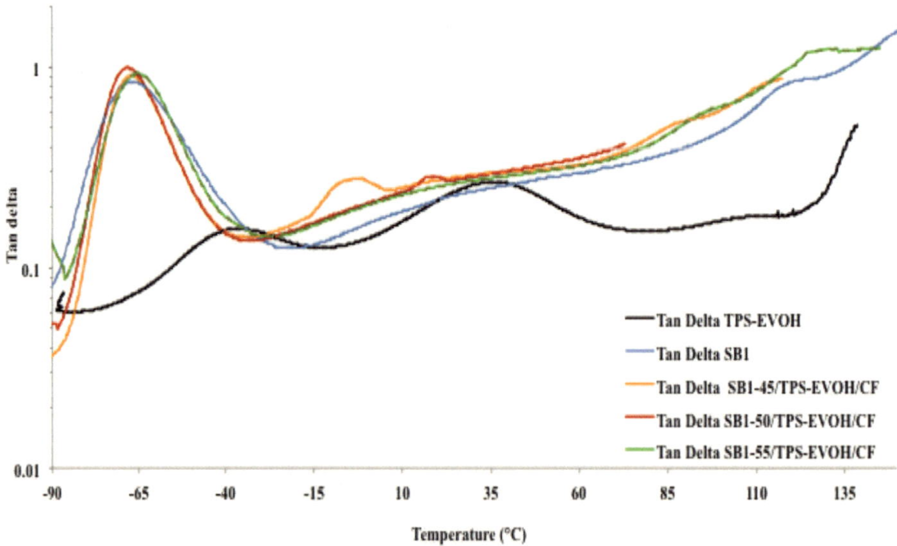

Figure 6. Tan δ curve from DMA for SB1/TPS-EVOH/CF composites.

behavior, but even so below the blend TPS-EVOH. In general, the same behavior was observed in the composites prepared with SB2 and SB3 elastomers. The modulus of TPS composites is typically higher than synthetic thermoplastics, indicating that inclusion of elastomer to TPS-EVOH has no reinforcement in matrix, due to the complexity of structure of the obtained material. Previous works are contradictory about the increasing and diminishing of storage modulus value about the reinforcement when particles are added to a polymer matrix [25].

Comparing TPS-EVOH/CF composites with SB1, SB2, and SB3, the storage modulus decreases, which has lower styrene content in elastomer; this behavior can be attributed to a steric hindrance of hard segments of styrene present in elastomer (**Figure 4**). The behavior might be due to free movement in the polymer chain at high temperatures and have some agreement with the results of DSC analysis (see **Table 2**).

The keratin addition has a significant effect on SB/TPS-EVOH/CF composites, increasing the initial value of storage modulus with respect to pure SB copolymer, causing a significant reinforcement effect in the rubber plateau region [31, 42]. A good interaction of particle aggregates in a polymer matrix can reflect in decreasing storage modulus when they reach the plateau region, also indicating that composites have better high-temperature stability, as can be observed for composite with lower elastomer content in composite (In general, SB1, SB2, and SB3/TPS-EVOH/CF composites present the similar behavior.) The Tg values of composites using tan δ (**Figure 6**) signal are, however, not affected by the CF inclusion due to the aggregate size and limited surface area [45]. The tan δ (**Figure 6**) is an useful tool to identify the interaction existing between the polymeric matrix (including the TPS-EVOH) and the keratin as reinforcement, in which a strong bond is reflected in a low tan δ value, although in this case, it has been used as an elastomeric matrix, which have higher tan δ values. It is observed that Tg value in the SB/TPS-EVOH/CF compounds is affected significantly with the CF addition and with the TPS-EVOH presence, and this is could be related to the result from tan δ curve. Nevertheless, the SB1/TPS-EVOH/CF composite, at −94°C, has the higher Tg value compared to the SB2/TPS-EVOH/CF and SB3/TPS-EVOH/CF composites. This same behavior has been reported before, in reference to no significant changes in Tg values by the effect of addition of CF as a reinforcement in the polymeric matrix [24, 32]. Jong [45] reports that for elastomer-protein composites, in the rubber plateau region, there is a very significant increase in the equilibrium storage modulus of composites of elastomers reinforced with proteins, and a better recovery behavior after eight cycles of dynamic strain, which indicates a stronger filler-rubber interaction. It was also found that the presence of aggregates causes an improvement in the effect of filler interaction with the matrix.

3.4. Thermogravimetric analysis (TGA)

The thermogravimetric analysis is used to predict the thermal stability of materials. When a reinforcement is added to a polymer matrix, it is necessary to identify the filler effect. For TPS, it has been reported that present thermal degradation due to loss of water below 140°C, around 200°C attributed to evaporation of water and glycerol and around 330°C from carbonization of starch [26, 38]. In the other hand, keratin decomposition has been reported before, presenting three steps, first around 210°C due to the protein denaturation, second one around 360°C resulting in a total destruction of protein, and third around 510°C for complete protein decomposition [2]. This is indicative that keratin is resistant to the action of elevated temperature. Other reports [1] found that CF lost moisture around 30 and 116°C, and temperature between 214 and 410°C presents the main decomposition stage with approximately 65% loss weight associated with breaking off the disulfide bonds in keratin structure, and the denaturation of the beta-protein structure as well as C—C bond degradation in the polymer backbone. **Figure 7** shows the TGA thermogram

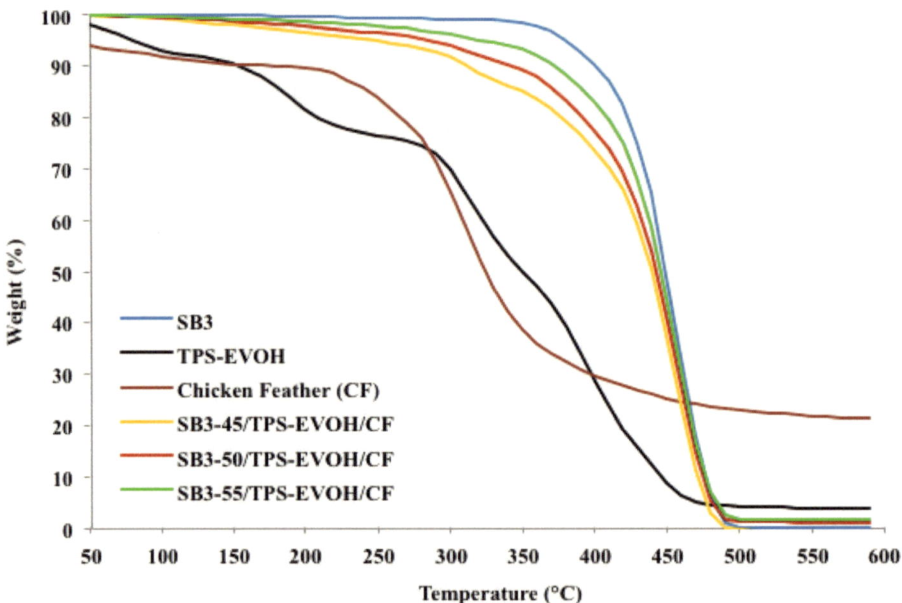

Figure 7. TGA thermograms for SB3, TPS-EVOH, CF, and SB3/TPS-EVOH/CF composite.

	Loss weight (%) at temperature (°C)						
Material	**200**	**250**	**300**	**350**	**400**	**450**	**500**
Chicken feather	89.7	83.8	65.3	38.6	29.7	25.3	23.0
TPS-EVOH	81.5	76.3	69.6	49.8	28.4	8.7	3.9
SB1	99.5	99.2	98.9	97.9	87.5	36.4	0.7
SB1-45/TPS-EVOH/CF	97.3	95.6	92.6	86.0	74.8	33.5	2.7
SB1-50/TPS-EVOH/CF	97.8	96.7	94.7	90.3	78.9	33.8	0.1
SB1-55/TPS-EVOH/CF	98.5	97.7	96.4	93.5	82.5	34.8	0.7
SB2	99.4	99.0	98.7	97.7	90.0	50.0	4.6
SB2-45/TPS-EVOH/CF	97.4	96.0	93.1	87.0	77.5	43.3	4.1
SB2-50/TPS-EVOH/CF	97.3	95.8	93.7	89.6	81.2	45.2	1.5
SB2-55/TPS-EVOH/CF	98.9	98.0	96.6	94.2	85.5	46.7	0.8
SB3	99.7	99.4	99.2	98.5	90.2	49.0	0.1
SB3-45/TPS-EVOH/CF	96.6	94.8	91.6	85.0	73.5	37.6	0.9
SB3-50/TPS-EVOH/CF	97.7	96.4	94.0	89.3	77.4	41.6	1.2
SB3-55/TPS-EVOH/CF	98.7	97.8	96.1	93.2	82.9	45.0	1.7

Table 3. Thermal data of loss weight for SB/TPS-EVOH/CF composites at different temperatures.

for SB3, CF, TPS-EVOH, and SB3/TPS-EVOH/CF composites with different SB3 contents. It can be observed that TPS-EVOH has three degradation steps, at around 130, 250, and 350°C, which are similar to reports of TPS thermal degradation [38]. CF has a similar behavior reported before as was discussed previously [1, 2]. Composite SB3/TPS-EVOH/CF has a similar behavior to that of SB3, so inclusion of elastomer improves the thermal degradation of TPS/EVOH/CF, due to the elastomer that has higher decomposition temperatures. A combinatorial effect of elastomer and CF can be present in the composites, as some reports found that keratin miscibility with polymer matrices increases the decomposition temperature in composites [25], in this work the content of CF was kept constant and the effect could be evaluated individually in thermal decomposition.

The composites with SB1 and SB2 have similar behavior compared with SB3 elastomer. **Table 3** shows the thermal data for SB/TPS-EVOH/CF composites at different temperatures with the aim to analyze. SB1/TPS0EVOH6CF composites present a lower loss weight when elastomer increases in formulation, which make sense, especially in region of 350–450°C, which is the main elastomer decomposition range. Composites prepared with the SB2 and SB3 elastomer has a better thermal resistance and better stability at higher temperatures, due to a lower loss weight, mainly at higher temperatures, and composites with SB3 elastomer showed a slight decrease in loss weight compared to composites with SB2 elastomer, which have the better thermal stability according to the results. Thermoplastic elastomer (TPE) is known due its higher thermal stability.

4. Conclusions

The infrared spectroscopy was not conclusive but indicates that there is no chemical interaction among components of composite. According with the styrene content present in the blend, it is possible to observed a better interaction among soft segments of the SB elastomers with the composites TPS-EVOH/CF. DMA results showed that the inclusion of SB elastomers in TPS-EVOH/CF do not have a positive effect, due to the decreased storage modulus compared to TPS-EVOH and pure SB, but the composites with lower SB content have a better behavior at lower and higher temperatures compared to the rest of the composites. The composites prepared with CF have been reported before that generates a similar behavior, but in this case, there was a synergetic effect, due to the structure of the composite material, so it can be attributed to SB elastomer of CF particles. The thermal behavior of the composites is very similar to that of the elastomer, which provides higher thermal stability as the styrene content increases therein.

Acknowledgements

Authors wish to thank Tecnológico Nacional de México (TNM) for their financial support for this research, code 6001.16-P. One of the authors (M.L.M.H.) wishes to thank CONACYT for the scholarship of postdoctorate program, number 291113-ITCM. Thanks also goes to Dynasol Elastomers S.A. de C.V. for providing SBS materials used in this research.

Author details

María Leonor Méndez-Hernández[1], José Luis Rivera-Armenta[1*], Zahida Sandoval-Arellano[2], Beatriz Adriana Salazar-Cruz[1] and María Yolanda Chavez-Cinco[1]

*Address all correspondence to: jlriveraarmenta@itcm.edu.mx

1 Research Centre on Petrochemical, Tecnológico Nacional de México-Instituto Tecnológico de Ciudad Madero, Altamira, Tamaulipas, Mexico

2 Faculty of Odontology, Universidad Autónoma de Coahuila Saltillo, Coahuila, Mexico

References

[1] Jimenez-Cervantes Amieva E, Velasco-Santos C, Martínez-Hernández AL, Rivera-Armenta JL, Mendoza-Martínez AM, Castaño VM. Composites from chicken feathers quill and recycled polypropylene. Journal of Composite Materials. 2015;**49**(3):275-283. DOI: 10.1177/0021998313518359

[2] Prochon M, Janowska G, Przepiorkowska A, Kucharska-Jastrzabek A. Thermal properties and combustibility of elastomer-protein composites. Journal of Thermal Analysis and Calorimetry. 2012;**109**:1563-1570. DOI: 10.1007/s10973-011-2028-1

[3] Martinez-Hernandez AL, Velasco-Santos C, De Icaza M, Castano VM. Microstructural characterisation of keratin fibres from chicken feathers. International Journal of Environment and Pollution. 2005;**23**(2):162-178. DOI: 10.1504/IJEP.2005.006858

[4] Tran CD, Prosenc F, Franko M, Benzi G. Synthesis, structure and antimicrobial property of green composites from cellulose, wool, hair and chicken feather. Carbohydrate Polymers. 2016;**151**:1269-1276. DOI: 10.1016/j.carbpol.2016.06.021

[5] Senoz E, Wool RP, McChalicher CW, Hong CK. Physical and chemical changes in feather keratin during pyrolysis. Polymer Degradation and Stability. 2012;**97**(3):297-307. DOI: 10.1016/j.polymdegradstab.2011.12.018

[6] Brebu M, Spiridon I. Thermal degradation of keratin waste. Journal of Analytical and Applied Pyrolysis. 2011;**91**(2):288-295. DOI: 10.1016/j.jaap.2011.03.003

[7] Hill P, Brantley H, Van Dyke M. Some properties of keratin biomaterials: Kerateines. Biomaterials. 2010;**31**(4):585-593. DOI: 10.1016/j.biomaterials.2009.09.076

[8] Winandy JE, Muehl JH, Micales JA, Raina A, Schmidt W. Potential of chicken feather fibre in wood MDF composites. Proceedings of EcoComp. 2003;**20**:1-6

[9] Janowska G, Kucharska-Jastrzabek A, Prochon M, Przepiorkowska A. Thermal properties and cmobustibility of elastomer-protein composites. Part II. Composites NBR-keratin. Journal of Thermal Analysis and Calorimetry. 2013;**113**:933-938. DOI: 10.1007/s10973-012-2796-2

[10] Martinez-Hernández AL, Velasco-Santos C, de-Icaza M, Castaoñ VM. Dynamical-mechanical and thermal analysis of polymeric composites reinforced with keratin biofibers from chicken feathers. Composites: Part B. 2007;**38**:405-410. DOI: 10.1016/j.compositesb.2006.06.013

[11] Yin XC, Li FY, He YF, Wang Y, Wang RM. Study on effective extraction of chicken feather keratins and their films for controlling drug release. Biomaterials Science. 2013;**1**:528-536. DOI: 10.1039/C3BM00158J

[12] Saravanan S, Sameera DK, Moorthi A, Selvamurugan N. Chitosan scaffolds containing chicken father keratin nanoparticles for bone tissue engineering. International Journal of Biological Macromolecules. 2013;**62**:481-486. DOI: 10.1016/j.ijbiomac.2013.09.034

[13] Kreskin Z, Urkmez AS, Hames EE. Novel keratin modified bacterial cellulose nanocomposite production and characterization for skin tissue engineering. Materials Science an d Engineering C. 2017;**75**:1144-1153. DOI: 10.1016/j.msec.2017.03.035

[14] Simmons S, Thomas EL. The use of transmission electron microscopy to study the blend morphology of starch/poly (ethylene-co-vinyl alcohol) thermoplastics. Polymer. 1998; **39**:5587-5599. DOI: 10.1016/S0032-3861(98)00025-1

[15] Rosa MF, Chiou B, Medeiros ES, Wood TG, Mattoso LHC, Orts WJ, Imam SH. Biodegradable composites based on starch/EVOH/glycerol blends and coconut fibers. Journal of Applied Polymer Science. 2008;**111**:612-618. DOI: 10.1002/app.29062

[16] Rosa MF, Chiou B, Medeiros ES, Wood DF, Williams TG, Mattoso LHC, Orts WJ, Imam SH. Effect of fiber treatments on tensile and thermal properties of starch/ethylene vinyl alcohol copolymers/coir biocomposites. Bioresource Technology. 2009;**100**:5196-5202. DOI: 10.1016/j.biortech.2009.03.085

[17] Vaz CM, Reis RL, Cunha AM. Use of coupling agents of enhance the interfacial interactions in starch-EVOH/hydroxyapatite. Biomaterials. 2002;**23**:629-635. DOI: 10.1016/S0142-9612(01)00150-8

[18] Leonor IB, Ito A, Onuma K, Kanzaki N, Reis RL. In vitro bioactivity of starch thermoplastic/hydroxyapatite composite biomaterials: An in situ study using atomic force microscopy. Biomaterials. 2003;**24**:579-585. DOI: 10.1016/S0142-9612(02)00371-X

[19] Reis RL, Cunha AM, Allan PS, Bevis MJ. Structure development and control of injection-molded hydroxyapatite-reinforced starch/EVOH composites. Advances in Polymer Technology;**16**:263-277

[20] Sinha Ray S. Polymer/layered silicate nanocomposite: A review from preparation to processing. Progress in Polymer Science. 2003;**28**:1539-1641. DOI: 10/1016/j.progpolymsci.2003.08.002

[21] Flores-Hernández CG, Colín-Cruz A, Velasco-Santos C, Castaño VM, Rivera-Armenta JL, Almendarez-Camarillo A, García-Casillas PE, Martínez-Hernández AL. All green composites from fully renewable biopolymers: Chitosan-starch reinforced with keratin from feathers. Polymer. 2014;**6**:686-705. DOI: 10.3390/polym6030686

[22] Barone JR. Polyethylene reinforced with keratin fibers obtained from chicken feathers. Composites Science and Technology. 2005;**65**:683-692. DOI: 10/1016/j.compscitech.2004.09.030

[23] Cheng S, Lau KT, Liu T, Zhao Y, Lam PM, Yin Y. Mechanical and thermal properties of chicken feather fiber/PLA green composites. Composites Part B: Engineering. 2009;**40**(7):650-654. DOI: 10.1016/j.compositesb.2009.04.011

[24] Barone JR, Schmidt WF. Polyethylene reinforced with keratin fibers obtained from chicken feathers. Composites Science and Technology. 2005;**65**:173-181. DOI: 10.1016/j.compscitech.2004.06.011

[25] Villarreal-Lucio DS, Rivera-Armenta JL, Rivas-Orta V, Díaz-Zavala NP, Páramo-García U, Gallardo-Rivas NV, Chávez-Cinco MY. Manufacturing composites from chicken feathers and polyvinyl chloride (PVC). In: Thakur VK, Thakur MK, Kessler MR, editors. Handbook of Composites from Renewable Materials: Design and Manufacturing. 2nd ed. Beverly: Wiley-Scrivener Publishing; 2017. pp. 159-174

[26] Shanks R, Kong I. Thermoplastic starch. In: El-Sonbati A, editor. Thermoplastic Elastomers. InTechOpen; 2012. pp. 95-116. DOI: 10.5772/36295

[27] Méndez-Hernández ML. Doctorate Thesis "Accelerated Aging Study of HDPE-Thermoplastic Starch Blends", Degree in Polymer Technology, Research Center in Applied Chemistry. Saltillo, Coahuila Mexico: CIQA; 2010

[28] Wang C, Guo Z, Fu S, Wu W, Zhu D. Polymers containing fullerene or carbon nanotube structures. Progress in Polymer Science. 2004;**29**:1079-1141. DOI: 10.1016/j.progpolymsci.2004.08.001

[29] Paul DR, Robeson LM. Polymer nanotechnology: Nanocomposites. Polymer. 2008;**49**:3187-3204. DOI: 10.1016/j.polymer.2008.04.017

[30] Wrzesniewska-Tosik K, Ademiec J. Biocomposites with a content of keratin from chicken feathers. Fibers & textiles in Eastern Europe. 2007;**15**:106-112

[31] Zha W, Han CD, Moon HC, Han SH, Lee DH, Kim JK. Exfoliation of organoclay nanocomposites based on polystyrene-block-polyisoprene-block-poly (2-vinylpyridine) copolymer: Solution blending versus melt blending. Polymer. 2010;**51**(4):936-952. DOI: 10.1016/j.polymer.2009.12.030

[32] Spitalskya Z. Carbon nanotube-polymer composites: Chemistry, processing, mechanical and electrical properties. Progress in Polymer Science. 2010;**35**(3):357-401. DOI: 10.1016/j.progpolymsci.2009.09.003

[33] Kviklys A, Lukosiute I. Kinetics and peculiarities of polyamide coatings formation from solutions. Materials Science (Medziagotyra). 2004;**10**:177-181

[34] Hernández-Martínez T, Salazar-Cruz BA, Rivera-Armenta JL, Chávez-Cinco MY, Méndez-Hernández ML, Páramo-García U. Evaluation of addition of reactive resin for an adhesive formulation of pressure-sensitive adhesive. In: Rudawska A, editor. Adhesives. Applications and Properties. InTechOpen; 2016. pp. 201-218. DOI: 10.5772/62603

[35] Salazar-Cruz BA, Rivera-Armenta JL, García-Alamilla R, Mendoza-Martínez AM, de la Garza AE, Espiricueto SM. Evaluación térmica del curado de adhesivos base SBR usando peróxido de dicumilo. Quimica Nova. 2015;**38**:651-656. DOI: 10.5935/0100-4042.20150067

[36] Méndez-Hernández ML, Tena-Salcido CS, Zandoval-Arellano Z, González-Cantú MC, Mondragón M, Rodríguez-González FJ. The effect of thermoplastic starch on the properties of HDPE/TPS blends during UV-accelerated aging. Polymer Bulletin. 2011;**67**:903-914. DOI: 10.1007/s00289-011-0501-4

[37] Rivera-Armenta JL, Flores-Hernández CG, Del Angel-Aldana RZ, Mendoza-Martínez AM, Velasco-Santos C, Martínez-Hernández AL. Evaluation of graft copolymerization of acrylic monomers onto natural polymers by means infrared spectroscopy. In: Theophanides T, editor. Infrared Spectroscopy: Materials Science, Engineering and Technology. InTechOpen; 2012. pp. 245-260. DOI: 10.5772/2055

[38] Mahdieh Z, Bagheri R, Eslami M, Amiri M, Shokrgozar MA, Mehrjoo M. Thermoplastic starch/ethylene vinyl alcohol/forsterite nanocomposites as a candidate material for bone tissue engineering. Materials Science and Engineering C. 2016;**C69**:301-310. DOI: 10.1016/j.msec.2016.06.43

[39] Ma B, Qiao X, Hou X, Yang Y. Pure keratin membrane and fibers from chicken feather. International Journal of Biological Macromolecules. 2016;**89**:614-621. DOI: 10.1016/j.ijbiomac.2016.04.039

[40] Khosa MA, Wu J, Ullah A. Chemical modification, characterization, and application of chicken feathers as novel biosorbents. RSC Advances. 2013;**3**(43):20800-20810. DOI: 10.1039/C3RA43787F

[41] Reddy N, Hu C, Yan K, Yang Y. Thermoplastic films from cyanoethylated chicken feathers. Materials Science and Engineering: C. 2011;**31**(8):1706-1710. DOI: 10.1016/j.msec.2011.07.022

[42] Morton M, McGrath JE, Juliano PC. Structure-property relationships for styrene-diene thermoplastic elastomers. Journal of Polymer Science, Polymer Symposia. 1969;**26**:99-115. DOI: 10.1002/polc.5070260107

[43] Taguet A, Huneault MA, Favis BD. Interface/morphology relationships in polymer blends with thermoplastic starch. Polymer. 2009;**50**:5733-5743. DOI: 10.1016/j.polymer.2009.09.055

[44] Jiang W, Qiao X, Sun K. Mechanical and thermal properties of thermoplastic acetylated starch/poly(ethylene-co-vinyl alcohol) blends. Carbohydrate Polymers. 2006;**65**:139-143. DOI: 10/.1016/j.carbpol.2005.12.038

[45] Jong L. Effect of soy protein concentrate in elastomer composites. Composites: Part A. 2006;**37**:438-446. DOI: 10.1016/j.compositesa.2005.05.042

Chapter 4

Production and Characterization of Starch Nanoparticles

Normane Mirele Chaves Da Silva,
Fernando Freitas de Lima,
Rosana Lopes Lima Fialho,
Elaine Christine de Magalhães Cabral Albuquerque,
José Ignacio Velasco and Farayde Matta Fakhouri

Additional information is available at the end of the chapter

http://dx.doi.org/10.5772/intechopen.74362

Abstract

In recent years, the increasing interest in nanomaterials of natural origin has led to several studies in the area of nano-sized particles from natural polysaccharide polymers, such as cellulose, starch, and chitin. These nanomaterials are used especially as a reinforcement in a polymeric matrix to improve the mechanical and barrier properties of the materials. Starch is a sustainable, abundant biopolymer produced by many plants as a source of storage energy; the main uses of starch are as food and industrial applications. However, recently their use as filler in polymeric matrix (nanoparticles) has attracted attention. Starch nanoparticles (SNPs) can be produced by many methods, using chemical, enzymatic, and physical treatments. The size distribution, crystalline structure, and physical properties of the SNPs may vary from one method to another. These nanoparticles are a very interesting alternatives not only for the polymeric filler but also for the renewability and biodegradability, since they show characteristics inherently of starch granules.

Keywords: methods, nanostarch, nanotechnology

1. Introduction

Nanotechnology is considered a study which involves science, medical, engineering, and technology at the nanoscale level; basically, nanotechnology involves the use of nanoparticles

ranging from 1 to 100 nm size [1]. In recent years, the use of nanotechnology for applications in the food industry has become more apparent, such as protection against biological and chemical deterioration, increasing bioavailability, enhancement of physical properties, and others [2]. However, the high cost of nanotechnology can make it difficult to its application in commercial scale. Therefore, the search for alternative materials and cheap to be used in the nanotechnology has been studied. Starch being a biodegradable natural polymer is a great alternative for the production of nanocrystals or nanoparticles. These materials can be produced by different methods, using chemical, enzymatic, and physical treatments and may be utilized as drug carriers, quality indicator for food products (nanoencapsulation), and reinforcement biodegradable and nonbiodegradable polymeric matrices [3].

2. Production and characterization methods of starch nanoparticles

The acid hydrolysis is the most commonly adopted method to produce starch nanocrystals (SNC). Usually, the starches submitted to this method have a two-step hydrolysis reaction: in the first step, a fast hydrolysis occurs, and in the second step, a slow hydrolysis occurs. For some authors there are three important steps of the acid hydrolysis: rapid, slow, and very slow [4–6]. In the first stage, the hydrolysis of the amorphous parts of the granules are attacked, while the slow step is the erosion of the crystalline regions [7]. Starch nanocrystals produced for this method present high crystallinity and a platelet-like shape. In this process, starch is diluted in acid (hydrochloric or sulfuric) maintained under constant stirring for a prolonged period with temperature control. After the hydrolysis period, the nanocrystals are differentiated from the acid by centrifugation and washed with distilled water until neutrality of the eluent. Finally, for a homogeneous dispersion of the nanocrystals, the suspension is submitted to a mechanical procedure (ULTRA-TURRAX).

The botanic origin of starch influences the thickness of SNC, which can vary between 4 nm for wheat starch and 8 nm for potato starch and crystalline organization and consequently the size of SNC. This is related to the fact that the acid hydrolysis occurs in the amylose molecules of the starch granules, and depending on the botanic origin of starch, it can be occurred in different sites of granule structure, for example, in the amorphous region (wheat starch), interspersed among amylopectin clusters in both the crystalline and amorphous regions (maize starch), and in bundles between amylopectin clusters or co-crystallized with amylopectin (potato starch) [3]. Thus, depending on the crystalline organization (amylose content), SNC can present larger sizes, since amylose is believed to jam the pathways for hydrolysis. Generally, SNC presents platelet-like morphology.

The lengthy duration of the acid hydrolysis (until 40 days) implies in a low yield (around 5%) [6]. Thus, this method is not appropriate for practical applications due to the long treatment period, its low yield, and use of concentrated acid, which can cause a negative impact on the environment.

The study carried out by Gonçalves et al. [8] using the acid hydrolysis method used to modify the starch extracted from the crude seeds of the pinhão (*Araucaria angustifolia*) was effective, resulting in nanometric particles. The starch modified by this method showed the greatest differences compared to common starch, being the most soluble, translucent, and hygroscopic among the samples. The authors conclude that the greater solubility and reduced turbidity are interesting from a commercial standpoint, showing that pinhão starch nanoparticles could be useful for the development of coating materials or films. In another study, the method showed an ability to form a strong elastic gel of starch nanocrystals [9]. Some recent studies by production of nanostarch by acid hydrolysis can be observed in **Table 1**.

Gamma radiation (γ-radiation) can be used to develop starch nanoparticles, since this technique can break large molecules into smaller fragments and is capable of cleaving glycosidic linkages. The technique consists in mixing starch and boiling water by stirring it for obtaining a homogenous paste, and then the suspension is irradiated using gamma ray, which generates active free radicals that are then responsible for the hydrolysis of starch. The fragmentation of starch results from the cleavage at the amorphous regions, instead of at the crystallite regions. In this sense, the gamma radiation method is very similar to that of the starch acid hydrolysis. Generally, the diameter of the nanoparticles obtained by this method is below 100 nm. Besides that, these nanoparticles also have nanocrystal aggregates, due to the large number of OH groups on their surface, which becomes strongly associated by hydrogen bonding, leading to fast thermal degradation [14]. The researches involving this method are still scarce and do not report the yield of process, which prevents the comparison with other methods.

Gamma radiation research demonstrates satisfactory results in the characterization and production of starch nanoparticles from cassava and waxy maize; the average sizes determined were (31 ± 5) nm and (41 ± 7) nm, respectively. The study shows that gamma radiation is a successful methodology to obtain starch nanoparticles able to be used as starch reinforcement and as a good alternative to production of starch nanoparticles, with low cost and using a simple and scalable methodology [14]. Gonzales Seligra et al. [28] also obtained average sizes less than 100 nm by producing starch nanoparticles by this same method. The insertion (0.6 wt%) of these nanoparticles in PBAT/TPS films improved the mechanical properties of the blend.

The production of starch nanoparticles (SNPs) using physical treatments is still recent, with high-pressure homogenization and ultrasound treatment being more utilized methods. In these methods, there are no chemical treatments or addition of chemical reagents. Some

Starch source	Time of reaction (days)	Size or size distribution (nm)	Yield (%)	Morphology	Reference
Amaranth	10	376	3.6	Lamellar structure	Sanchez de la Concha et al. [10]
Waxy maize	5	58	—	Platelet	Bel Haaj et al. [11]
Cassava	5	47–178	30	Spherical	Costa et al. [12]
Amadumbe	5	180–280	25	Platelet	Mukurumbira et al. [13]

Table 1. Starch source, time of reaction (days), size or size distribution (nm), yield (%), and morphology found in different studies on obtaining nanostarchs by acid hydrolysis.

advantages of these methods are that they are simple, effective, and environmentally friendly. Besides that, they might reduce the processing time to generate SNPs, increase the yield in NP production, and avoid various purification steps such as the acid hydrolysis [15]. In this context, ultrasound treatment shows up as a viable alternative. The method consists in sonication a starch suspension (starch and water) with controlled temperature for a fixed time, using ultrasound equipment. During the ultrasonication occurs a transfer the energy for to starch particles by cavitation, which is the collapse of microbubbles that burst and propagate as a sound wave through the solution. So microjets are formed with high velocities resulting the shear forces which may break covalent bonds of the starch and reduce the particle size. The ultrasonication process influences the crystalline structure of the starch (amylopectin), leading to nanoparticles with low crystallinity or an amorphous character. The ultrasonication in the starch can be affected by many factors such as ultrasonication power and frequency, time, and treatment temperature, besides the characteristics of starch dispersions, which are the concentration and botanical origin of starch and the dispersion solvent. Thus, the starch nanoparticles obtained for this method also can vary; for example, the size of the nanoparticles may vary between 30 nm and 200 nm. Some studies can be observed in **Table 2**.

The high-pressure homogenization is commonly in the chemical, pharmaceutical, food, and biotechnology industries. In this treatment, changes not only in the products but also in the particles, colloids, or macromolecules which are product constituents may occur. Thus, novel applications for the high-pressure homogenization are researched [17] between the productions of starch nanoparticles. The method consists of the manipulation of a continuous flow of liquid through microfabricated channels; the starch slurry is passed by a microfluidizer, which can be intensified by external pressure sources, external mechanical pumps, integrated mechanical micropumps, or electrokinetic mechanisms, which result in the breakage of the hydrogen bonding inside the large particles by the mechanical shear forces. Homogenization pressures can reach up to 350 MPa. The nanoparticles obtained by this method may vary by 10 nm in size. In this method, the partial or complete destruction of the crystalline structure can also occur, and only a low concentration of starch slurry could be processed for homogenization.

Recently, studies involving the ultrasound treatment and high-pressure homogenization methods show the development of nanoparticles of starch with nanometric scale sizes and the ability to form films [8, 18, 19]. The use of the ultrasound treatment method in pinhão

Starch source	Concentration (wt%)	Power (W)	Size or size distribution (nm)	Morphology	Reference
Waxy maize	2.0	400	40.0	Platelet	Boufi et al. [15]
Cassava	1.5	50	75.51	Spherical	da Silva et al. [16]
Pinhão	20.0	100	454.3	Concave	Gonçalves et al. [8]
Potato	3.0	100	77.0	Spherical	Chang et al. [18]
Waxy maize	1.5	136	100–200	—	Bel Haaj et al. [20]

Table 2. Starch source, concentration (wt%), power (W), and size or size distribution (nm) found in different studies on obtaining nanostarchs by ultrasound treatment.

(*Araucaria angustifolia*) seeds can be useful for the development of novel biocomposites, with improved properties to be employed such as coating materials or films [8]. In other researches, the study not only shows an easily controllable methodology to prepare starch nanoparticles of small size and narrow distribution through precipitation but also provides an approach to produce starch nanoparticles with high efficiency and low cost, decreasing in viscosity of starch aqueous paste and not requiring any chemical treatment [18, 20].

The high-pressure homogenization researches have shown results of starch nanoparticles analyzed by transmission electron microscopy (TEM) and dynamic light scattering (DLS), which showed that starch nanoparticles had narrow size distribution, high dispersibility, and spherical shape [21] and films obtained from high-pressure homogenized dispersions had good moisture barrier capacity, better film transparency, and higher tensile strength but, however, lower elongation [19].

The utilization of the starch nanoparticles as polymeric filler is recent; however, the researchers showed satisfactory results. The nanoparticle presents at least one of its dimensions which is lower than 100 nm; thus, this nanometric dimension may result in a better dispersion and compactness of polymeric structure. The insertion of starch nanoparticle in a polymer results a new material, known as nanocomposite, with great properties that are not seen in traditional composites.

The incorporation of SNPs improves the mechanical properties and water vapor permeability (WVP) and also the biodegradability of the composites. The decrease in the WVP of nanocomposite films is attributed to compactness of the polymeric structure which is resulting in water vapor diffusion more difficult and consequently reducing permeability; in case the reinforcement is derived from the same material as the matrix, such as starch nanocrystals dispersed in starch films, it could have better compatibilization, since they show characteristics inherently from starch granules. It is worth pointing out that the concentration of starch nanoparticle inserted in a matrix polymeric must be carefully analyzed, once a lower nanoparticle concentration leads to better dispersion in the films and less clustering that hinders the passage of water and reduces permeability. On the other hand, the increase in WVP with a higher concentration of the SNPs can be related with a more nanoparticle grouping allowing diffusion of water molecules. Recently, some studies involving the use of SNP as filler in starch films showed that when the concentration of SNPs inserted was less than 6%, the WVP presented a reduction. However, these same studies showed the use of a higher concentration of the SNPs resulting in an increase in WVP [22, 23]. The increase in WVP is not feasible for food packaging use; it offers increased food degradation rate. The WVP values are essential for the possible packaging application use of the biofilms. The material that where very permeable to water vapor may be suitable for the packaging of fresh food, whereas a slightly permeable biofilm may be useful for the packaging of dehydrated food [24].

The mechanical properties are proven be one the most important parameters for biofilm analysis, which usually presented poor mechanical properties. One alternative for improving these properties is the use of starch nanoparticle as reinforcement agent. For the mechanical properties, also the nanometric dimension of the starch nanoparticles can result in strong interactions with different matrices, once they have the capacity of occupying inter- or intramolecular space, resulting in densification of the film [25]. Besides that, the nanoparticle presents high specific surface area which can result in a better between filler and polymeric matrix interfacial interaction, which result in an increase of the nanocomposite strength [26]. So, the polymeric film incorporated with the SNP can present an increase in the tensile strength and modulus of elasticity, associated with decrease in the elongation percentage. These effects are heavily dependent on the SNP concentrations incorporated in the nanocomposites, because high concentrations of SNPs when incorporated in a matrix can cause aggregation, which leads to weaken the interface adhesion between the nanoparticle and matrix [27]. Li et al. [22] studied the incorporation of starch nanocrystals (SNC) in pea starch films. The authors concluded the concentrations bigger than 5% of SNC when incorporated in the films resulted in a decreased tensile strength associated with increase in elastic modulus.

Besides that, SNPs can speed up the biodegradation process. The influence of the SNP in the faster biodegradability of the composites is due to the fact that in soil, water diffuses into the polymer sample, causing swelling and enhancing biodegradation due to increases in microbial growths [28]. Costa et al. [12] studied the use of cassava starch nanocrystals (CSN) obtained by acid hydrolysis to strengthen nanocomposite films from the same matrix. The authors conclude that the large percentage of loss of film mass was found over the biodegradability test and the film with 10% CSN showed a larger weight loss, which is probably associated with greater microorganism access.

Acknowledgements

The authors thank the Conselho Nacional de Desenvolvimento Científico e Tecnológico (CNPq) and the Coordenação de Aperfeiçoamento de Pessoal de Nível Superior (CAPES).

Conflict of interest

The authors declare no conflicts of interest.

Author details

Normane Mirele Chaves Da Silva[1], Fernando Freitas de Lima[2], Rosana Lopes Lima Fialho[1], Elaine Christine de Magalhães Cabral Albuquerque[1], José Ignacio Velasco[3] and Farayde Matta Fakhouri[3,4]*

*Address all correspondence to: farayde@gmail.com

1 Polytechnic School, Federal University of Bahia, Salvador, BA, Brazil

2 School of Chemical Engineering, University of Campinas, Campinas, SP, Brazil

3 ESEIAAT, Universitat Politècnica de Catalunya - Barcelona TECH, Barcelona, Spain

4 Faculty of Engineering, Federal University of Grande Dourados, Dourados, MS, Brazil

References

[1] Bera A, Belhaj H. Application of nanotechnology by means of nanoparticles and nano-dispersions in oil recovery—A comprehensive review. Journal of Natural Gas Science and Engineering. 2016;**34**:1284-1309

[2] He X, Hwang HM. Nanotechnology in food science: Functionality, applicability, and safety assessment. Journal of Food and Drug Analysis. 2016;**24**:671-681

[3] Le Corre D, Angellier-Coussy H. Preparation and application of starch nanoparticles for nanocomposites: A review. Reactive and Functional Polymers. 2014;**85**:97-120

[4] Putaux JL, Molina-Boisseau S, Momaur T, et al. Platelet nanocrystals resulting from the disruption of waxy maize starch granules by acid hydrolysis. Biomacromolecules. 2003;**4**:1198-1202

[5] Angellier H, Putaux JL, Molina-Boisseau S, et al. Starch nanocrystal fillers in an acrylic polymer matrix. In: Macromolecular Symposia. 2005. pp. 95-104

[6] Le Corre D, Bras J, Dufresne A. Starch nanoparticles: A review. Biomacromolecules. 2010;**11**:1139-1153

[7] Kim HY, Park SS, Lim ST. Preparation, characterization and utilization of starch nanoparticles. Colloids and Surfaces B: Biointerfaces. 2015;**126**:607-620

[8] Gonçalves PM, Noreña CPZ, da Silveira NP, et al. Characterization of starch nanoparticles obtained from Araucaria angustifolia seeds by acid hydrolysis and ultrasound. LWT—Food Science and Technology. 2014;**58**:21-27

[9] Jiang S, Liu C, Han Z, et al. Evaluation of rheological behavior of starch nanocrystals by acid hydrolysis and starch nanoparticles by self-assembly: A comparative study. Food Hydrocolloids. 2016;**52**:914-922

[10] de la Concha BBS, Agama-Acevedo E, Nuñez-Santiago MC, et al. Acid hydrolysis of waxy starches with different granule size for nanocrystal production. Journal of Cereal Science. 2018;**79**:193-200

[11] Bel Haaj S, Thielemans W, Magnin A, et al. Starch nanocrystals and starch nanoparticles from waxy maize as nanoreinforcement: A comparative study. Carbohydrate Polymers. 2016;**143**:310-317

[12] Costa ÉK de C, de Souza CO, da Silva JBA, et al. Hydrolysis of part of cassava starch into nanocrystals leads to increased reinforcement of nanocomposite films. Journal of Applied Polymer Science; 134. Epub ahead of print 2017. DOI: 10.1002/app.45311

[13] Mukurumbira A, Mariano M, Dufresne A, et al. Microstructure, thermal properties and crystallinity of amadumbe starch nanocrystals. International Journal of Biological Macromolecules. 2017;**102**:241-247

[14] Lamanna M, Morales NJ, Garcia NL, et al. Development and characterization of starch nanoparticles by gamma radiation: Potential application as starch matrix filler. Carbohydrate Polymers. 2013;**97**:90-97

[15] Boufi S, Bel Haaj S, Magnin A, et al. Ultrasonic assisted production of starch nanoparticles: Structural characterization and mechanism of disintegration. Ultrasonics Sonochemistry. 2018;**41**:327-336

[16] da Silva NMC, Correia PRC, Druzian JI, et al. PBAT/TPS composite films reinforced with starch nanoparticles produced by ultrasound. International Journal of Polymer Science. 2017;1-10

[17] Wei B, Cai C, Xu B, et al. Disruption and molecule degradation of waxy maize starch granules during high pressure homogenization process. Food Chemistry. 2018;**240**:165-173

[18] Chang Y, Yan X, Wang Q, et al. High efficiency and low cost preparation of size controlled starch nanoparticles through ultrasonic treatment and precipitation. Food Chemistry. 2017;**227**:369-375

[19] Fu ZQ, Wang LJ, Li D, et al. Effects of high-pressure homogenization on the properties of starch-plasticizer dispersions and their films. Carbohydrate Polymers. 2011;**86**:202-207

[20] Bel Haaj S, Magnin A, Pétrier C, et al. Starch nanoparticles formation via high power ultrasonication. Carbohydrate Polymers. 2013;**92**:1625-1632

[21] Shi AM, Li D, Wang LJ, et al. Preparation of starch-based nanoparticles through high-pressure homogenization and miniemulsion cross-linking: Influence of various process parameters on particle size and stability. Carbohydrate Polymers. 2011;**83**:1604-1610

[22] Li X, Qiu C, Ji N, et al. Mechanical, barrier and morphological properties of starch nanocrystals-reinforced pea starch films. Carbohydrate Polymers. 2015;**121**:155-162

[23] Jiang S, Liu C, Wang X, et al. Physicochemical properties of starch nanocomposite films enhanced by self-assembled potato starch nanoparticles. LWT—Food Science and Technology. 2016;**69**:251-257

[24] Pagno CH, Costa TMH, De Menezes EW, et al. Development of active biofilms of quinoa (Chenopodium quinoa W.) starch containing gold nanoparticles and evaluation of antimicrobial activity. Food Chemistry. 2015;**173**:755-762

[25] Shi Y, Jiang S, Zhou K, et al. Influence of g-C3N4 nanosheets on thermal stability and mechanical properties of biopolymer electrolyte nanocomposite films: A novel investigation. ACS Applied Materials & Interfaces. 2014;**6**:429-437

[26] Wetzel B, Haupert F, Zhang MQ. Epoxy nanocomposites with high mechanical and tribological performance. Composites Science and Technology. 2003;**63**:2055-2067

[27] Dai L, Qiu C, Xiong L, et al. Characterisation of corn starch-based films reinforced with taro starch nanoparticles. Food Chemistry. 2015;**174**:82-88

[28] González Seligra P, Eloy Moura L, Famá L, et al. Influence of incorporation of starch nanoparticles in PBAT/TPS composite films. Polymer International. 2016;**65**:938-945

Chapter 5

Aspects on Starches Modified by Ionizing Radiation Processing

Mirela Brașoveanu and Monica-Roxana Nemțanu

Additional information is available at the end of the chapter

http://dx.doi.org/10.5772/intechopen.71626

Abstract

Starch is one of the most studied natural polymers due to its widespread and applications as well as to the global interest regarding renewable, cheap, and easy to process resources. The native form of starch is frequently subjected to different processing methods in order to modify its structure and thus to obtain some functional properties suitable in specific industrial applications. Radiation-based method is a "green tool" for modification of natural polymers, such as starch, cellulose, pectin, and chitosan, alginate, having advantages over conventional methods that involve chemical agents associated with environmental toxicity. Radiation processing of natural polymers involves a simple, ecofriendly, and fast process that has harmless feature and provides advanced materials with unique properties. The chapter intends to be an overview of the major findings in the last decade concerning the starches modified by ionizing radiation processing. Therefore, aspects strongly related to changes in physicochemical, functional, and structural properties of starches from various botanical origins are approached. The main key points of this topic are highlighted by a critical evaluation of the mentioned aspects and future perspectives are suggested.

Keywords: starch, modification, gamma radiation, electron beam, processing

1. Introduction

Starch is one of the most widespread and used natural polymers in different applications such as food, pharmaceutical and cosmetics, paper, and textile industries. Even starch is a renewable, cheap, and easy to process resources, its use as a native form has been restricted by some limitations (high viscosity, low solubility in cold water, paste instability, etc.) in specific applications due to its structure. Therefore, starch is frequently subjected to different processing methods (chemical, physical and enzymatic treatments) in order to modify its structure

resulting in functional properties suitable in specific industrial applications. The most used way to obtain modified starch is by chemical methods, which are most of the time complex, expensive, and time consuming.

Progressive methods of starch modification are generally considered the physical techniques (e.g., ionizing and non-ionizing radiation treatments, plasma treatment), which are fast, low cost, and environmentally friendly because they do not use polluting agents, do not allow the penetration of any toxic substances in the treated products, do not generate undesirable residual products, and do not require catalysts and laborious preparation of samples.

In the last decade, there is a great amount of reported studies related to the effects of ionizing radiation (gamma radiation, electron beam) on different type of starches. The studies concerning the impact of gamma radiation or electron beam (e-beam) were performed on starches extracted from various vegetal sources—cereals, tubers, legumes, and stems—as presented in **Table 1**.

The reported data showed that ionizing radiation processing generates free radicals on starch molecules that can alter their size and structure, leading to changes of functional and physicochemical properties of starch as a function of experimental processing parameters (irradiation dose, irradiation dose rate, moisture content of samples, and type of gas atmosphere).

Type of starch	Type of ionizing radiation/processing parameters	Investigated properties	Techniques of characterization	References
Cereal				
Corn	Gamma rays; 1–50 kGy, 1 kGy/h, in air	Apparent amylose content Thermal properties Pasting profile Granule morphology and crystallinity	Chemical methods, DSC, viscometry, X-ray diffraction, SEM	[1]
	Gamma rays; 3–50 kGy, 19 Gy/min, in air	Pasting behavior Thermal properties Spectral characteristics Granule morphology and crystallinity	Viscometry, DSC, FTIR, X-ray diffraction, PLM	[2]
	E-beam 6 MeV; 10–50 kGy, 2 kGy/min, in air	Apparent viscosity Pasting properties Thermal properties Colorimetric properties Molecular weight and molecular weight distribution Granule morphology Spectral characteristics	Rheology, viscography, DSC, UV-Vis spectrometry, GPC, SEM, FTIR	[3–6]
	E-beam 6 MeV; 1–10 kGy, in air	Colorimetric properties	UV-Vis spectrometry	[7]
	E-beam 6 MeV; 5–100 kGy, solid and liquid	Apparent viscosity, Molecular weight and radius of gyration Granule morphology	Viscometry, rheology, MALLS, SEM	[8]

Type of starch	Type of ionizing radiation/processing parameters	Investigated properties	Techniques of characterization	References
Wheat	Gamma rays; 0.5–10 kGy, in air, moisture content 11%	Proximate composition Color Swelling and solubility indices Light transmittance Syneresis Freeze thaw stability Total amylose content Water and oil absorption capacities Pasting parameters Spectral characteristics	Chemical and visible spectroscopic methods, viscography, FTIR	[9]
	Gamma rays; 3–50 kGy, 13.84 Gy/min, in air, moisture content 13.4%	Structural, spectral and morphological properties Thermal properties Apparent amylose content Water solubility index and swelling power Pasting behavior	EPR, FTIR, XRD, SEM, DSC Chemical methods, viscography and viscometry	[10]
	E-beam 6 MeV; 10–50 kGy, 2 kGy/min, in air	Apparent viscosity Intrinsic viscosity Thermal properties Colorimetric properties Molecular weight and molecular weight distribution	Rheology, viscometry, DSC, UV-Vis spectrometry, GPC	[4, 5, 11]
	E-beam 2 MeV; 1–4.4 kGy, relative humidity 70 ± 5%	Pasting properties Thermal properties Molecular weight distribution Microscopic characteristics	Viscography, DSC, HPSEC, SEM	[12]
	E-beam 3 MeV; 5–30 kGy	Pasting profile Swelling volume Leaching of carbohydrates and amylose	Viscography, chemical and spectroscopic methods	[13]
	E-beam/6 MeV, 1–10 kGy, in air	Colorimetric properties	UV-Vis spectrometry	[7]
Rice	Gamma rays; 1–5 kGy, 0.4 kGy/h, in air, moisture content 9%	Amylose and carboxyl contents Acidity Thermal properties Molecular weight distribution Granule morphology and crystallinity	Chemical methods, DSC, GPC, SEM, X-ray diffraction	[14]
	Gamma rays; 1–10 kGy, 0.4 kGy/h, in air, moisture content 12%	Swelling power and solubility Carboxyl content Pasting properties Thermal properties Granule morphology and crystallinity Spectral characteristics	Chemical methods, viscography, DSC, SEM, X-ray diffraction, FTIR	[15]
	E-beam 6 MeV; 10–50 kGy, 2 kGy/min, in air	Apparent viscosity Thermal properties Colorimetric properties Molecular weight and molecular weight distribution	Rheology, DSC, UV-Vis spectrometry, GPC	[4, 5]

Type of starch	Type of ionizing radiation/processing parameters	Investigated properties	Techniques of characterization	References
Tuber				
Potato	Gamma rays; 5–20 kGy, 2 kGy/h, in air, moisture content 10%	Carboxyl content Apparent amylose and amylose leaching Swelling power and solubility Syneresis Pasting properties Granule morphology and crystallinity	Chemical methods, spectroscopy, viscography, SEM, X-ray diffractometry	[16]
	Gamma rays; 10 and 50 kGy, 2 kGy/h, in air, moisture content 10%	Carboxyl content Swelling factor and amylose leaching Apparent amylose content Pasting properties Thermal properties Granule morphology and crystallinity Spectral characteristics In vitro digestibility	Chemical methods, DSC, viscography, FTIR, SEM, optical microscopy, X-ray diffraction	[17]
	E-beam 6 MeV; 10–50 kGy, 2 kGy/min, in air	Apparent viscosity Thermal properties Colorimetric properties Molecular weight and molecular weight distribution	Rheology, DSC, UV–Vis spectrometry, GPC	[4, 5]
	E-beam 6–7 MeV; 110–440 kGy, in air, moisture content 12%	Phase structure Structure morphology Dynamic viscosity Amount of carboxylic and carbonyl groups Solubility in cold water	Wide-angle X-ray diffraction, SEM, FTIR, rheology, chemical methods	[18]
Tapioca	E-beam 3 MeV; 5–30 kGy	Pasting profile Swelling volume Leaching of carbohydrates and amylose	Viscography, chemical and spectroscopic methods	[13]
Elephant foot yam (*Amorphophallus paeoniifolius*)	Gamma rays; 5–25 kGy, 2 kGy/h, in air	Color, pH Apparent amylose and carboxyl contents Swelling power and solubility Water absorption capacity Light transmittance Syneresis Pasting parameters Morphological characteristics and crystallinity Spectral characteristics Thermal analysis	Chemical and spectroscopic methods, viscography, SEM, FTIR, DSC	[19]

Type of starch	Type of ionizing radiation/processing parameters	Investigated properties	Techniques of characterization	References
Legume				
Bean	Gamma rays; 5–25 kGy, 185 Gy/h, in air	Solubility and swelling power Carboxyl content, pH Retrogradation Apparent amylose content and amylose leaching Water absorption capacity Pasting parameters Thermal properties In vitro digestability Granule morphology and crystallinity Antioxidant activity	Chemical methods, viscography, DSC, SEM, X-ray diffractometry	[20]
	Gamma rays; 5–15 kGy, 83 Gy/min	Color Apparent amylose and carboxyl contents Water and oil absorption capacities Bulk density Swelling and solubility indices Light transmittance Syneresis Freeze thaw stability Pasting properties Granule morphology and crystallinity Spectral characteristics	Chemical methods, viscography, visible spectroscopy, SEM, X-ray diffraction, FTIR	[21]
	Gamma rays; 5–20 kGy, 2 kGy/h, in air, moisture content 10%	Carboxyl content, pH Apparent amylose Swelling power and solubility Water absorption capacity Light transmittance Syneresis Pasting properties Granule morphology and crystallinity	Chemical methods, viscography, visible spectroscopy, SEM, X-ray diffractometry	[22]
	Gamma rays; 10 and 50 kGy, 2 kGy/h, in air, moisture content 10%	Carboxyl content Swelling factor and amylose leaching Apparent amylose content Pasting properties Thermal properties Granule morphology and crystallinity Spectral characteristics In vitro digestibility	Chemical methods, DSC, viscography, FTIR, SEM, optical microscopy, X-ray diffraction	[17]

Type of starch	Type of ionizing radiation/processing parameters	Investigated properties	Techniques of characterization	References
Stem				
Lotus	Gamma rays; 5–20 kGy, 2 kGy/h, in air, moisture content 12%	Apparent amylose content Carboxyl content, pH, Amylose leaching and swelling power Water absorption capacity Syneresis Pasting properties Granule morphology and crystallinity	Chemical methods, viscography, SEM, X-ray diffractometry	[23]
Sago	E-beam 3 MeV; 5–30 kGy	Pasting profile Swelling volume Leaching of carbohydrates and amylose	Viscography, chemical and spectroscopic methods	[13]

Table 1. Studies relevant to effects of ionizing radiation on various starches.

The chapter gives an overview of the major findings in the last decade concerning the starches modified by ionizing radiation processing. Therefore, aspects strongly related to changes in physicochemical, functional, and structural properties of starches from various botanical origins are approached. The main key points of this topic are highlighted by a critical evaluation of the mentioned aspects and future perspectives are suggested.

It is to be mentioned herein that in the last decade two other reviews regarding the impact of radiation processing on starch [24, 25] have been published, and the most recent one [25] has summarized only the gamma radiation influence on starches.

2. Fundamentals of radiation processing

Radiation is generally a form of energy characterized by its ability to move from one location to another, and it can be divided into non-ionizing (ultraviolet light, visible light, infrared radiation, microwaves, etc.) and ionizing (X-rays, gamma rays, electron beams, etc.) ones [26]. Ionizing radiation—mainly gamma radiation and electron beam—is the most used to modify starch macromolecules and, further, the approach will be made in context of the processing procedures using ionizing radiation. Gamma rays—electromagnetic radiation—are emitted by radionuclides such as isotopes of cobalt-60 (^{60}Co) and cesium-137 (^{137}Cs). Electron beam consists of accelerated electrons, which are charged particles generated from regular electricity using linear accelerators, and do not involve radioactive isotope sources. It is noteworthy that electron beam irradiation is similar to gamma processing with basically the same interaction with materials to be subjected to irradiation processing [27].

2.1. Interaction of radiation with matter

Gamma rays transfer energy by the photoelectric effect, Compton effect, and pairs generating, leading to liberation of fast electrons that lose energy by the same effects as accelerated electrons of the electron beams. Later on, the energy is absorbed by matter when electrons pass through it and two distinct primary effects, *ionization* and *excitation* of atoms and molecules of the substance occur as presented synthetically in **Figure 1**.

Ionization is the primary process by which a neutral atom or molecule becomes charged, and the resulting product is called *ion*. An ion formed by the loss or capture of an electron contains an unpaired electron, being actually a *free radical*, which is highly reactive chemical specie. The ejected electron may ionize further other atoms and molecules by successive collisions and ionizations. Excitation is another primary effect occurring when a high-energy charged particle passes through atoms imparting energy to atomic electrons without ejecting them and leads to an excited atom.

The secondary effects consist in different reactions of primary species (ions, excited molecules, or free radicals) that lead to the final products. These effects could be dissociation of an excited molecule into two radicals or into two different molecules. The free radicals participate further in recombination processes between themselves (radical-radical recombination) leading either to the initial molecule or new molecules. The formed radicals can also suffer a recombination with a new molecule abstracting a hydrogen atom forming a new free radical and a new molecule.

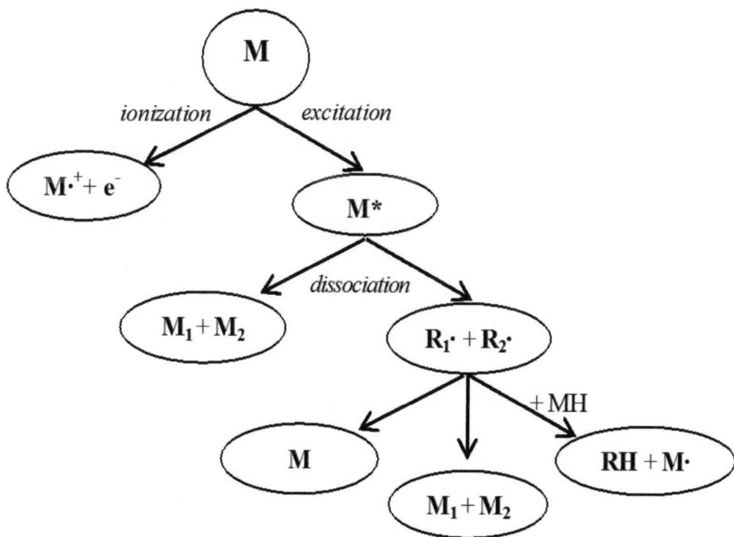

Figure 1. Fundamental processes of electrons passing through matter.

2.2. Radiation quantities

The most used dosimetric quantities and their units are the absorbed dose and the absorbed dose rate according to the International Commission on Radiation Units and Measurements (ICRU) Report no. 33 [28]. Moreover, these radiation quantities are the most important in the physical quantities used in the dosimetry field in order to optimize and control the irradiation process [29, 30].

The *absorbed dose, D,* is the amount of energy absorbed per unit mass of irradiated matter at a point in the region of interest [29]:

$$D = \frac{d\bar{\varepsilon}}{dm} \tag{1}$$

where $d\bar{\varepsilon}$ is the mean energy imparted by ionizing radiation to the matter in a volume element and dm is the mass of that volume element.

The SI derived unit of absorbed dose is the *gray* (Gy), which replaced the earlier unit of absorbed dose, the *rad* that is still tolerated, but less used:

$$1\,\text{Gy} = 1\ \text{J/kg} = 100\,\text{rad}$$

For this absorbed dose, the rate of change of it with time can be defined as the absorbed dose rate, $\dot{D}$:

$$\dot{D} = \frac{dD}{dt} \tag{2}$$

The SI derived unit of absorbed dose rate is the *Gy/s*.

The absorbed dose and the absorbed dose rate will be hereinafter referred as irradiation dose and irradiation dose rate, respectively.

2.3. Advantages and disadvantages of ionizing radiation processing

Ionizing radiation has many advantages in material processing, being an effective tool to induce changes in structure and functional properties of materials without environmental negative implication. Thus, it is an environmentally friendly process that involves no use of polluting agents, no generation of undesirable residual products, and no penetration of toxic substances in treated products. Despite the advantages, ionizing radiation processing is accompanied by some disadvantages as shown in **Table 2**.

Type of radiation	Advantages	Disadvantages
Gamma rays	-Cold method -High penetration depth	-Low dose rate -Long time processing (hours) -Use of radioactive source
Electron beam	-Cold method -High dose rate -Very short time processing (s) -Radiation can be turn on and off	-Low penetration depth

Table 2. Ionizing radiation processing: Advantages and disadvantages.

http://dx.doi.org/10.5772/intechopen.71626

3. Effects of ionizing radiation on functional properties

The physicochemical and functional properties of starch in various applications are of great interest, especially for manufacturers of starch-based products. Aspects related to the ionizing radiation effects on physical, chemical, and functional characteristics of starch are onward presented.

3.1. Physicochemical properties

The moisture contents of starches extracted from various botanical sources (lotus, sago, tapioca, wheat) were insignificantly affected by gamma radiation up to 20 kGy (dose rate ≤ 9 kGy/h) and electron beam up to 30 kGy [13, 23]. On the other hand, the moisture contents of rice starches were also insignificantly affected by gamma radiation at low doses (<1.5 kGy with dose rate of 0.63 kGy/h) [31], whereas a significant reduction in moisture content occurred at irradiation doses >2 kGy (dose rate of 0.4 kGy/h) as a result of radiation energy dissipation while ionizing radiation penetrates the starch sample [15]. Also, for starch extracted from elephant foot yam, the amount of moisture decreased significantly by gamma irradiation up to 25 kGy with dose rate of 2 kGy/h [19]. According to Reddy et al. [19], the reduction in the moisture content of starch sample by radiation processing may improve the shelf life of starch by avoiding the microorganisms' development.

pH of aqueous starch solutions decreased with increase of irradiation dose regardless of the botanical origin of the starch [5, 14, 16, 19, 23, 31–33]. The descending change of solution pH after irradiation could be attributed to the formation of chemical groups with acidic character such as carboxyl, carbonyl, or peroxide groups. Moreover, this behavior is sustained due to the fact that radiation processing of starch was generally performed in the presence of oxygen, thus promoting the appearance of free radicals, compounds with carbonyl bonds (aldehydes/ketones), organic peroxides, or other polysaccharide degradation products [34] that can lead to the increase of starch acidity. Therefore, the reduction of solution pH is strongly correlated with the increase of carboxyl content by the ionizing radiation processing of starch.

The water solubility can be improved concomitantly with the reduction of swelling power of granule by ionizing radiation processing for all starches. Therefore, the solubility value increased with the increase of irradiation dose for starches extracted from various botanical sources (corn, wheat, rice, potato, bean, elephant foot yam, lotus, chickpea, and Indian horse chestnut) [5, 9, 10, 15, 16, 18, 19, 21–23, 31–33, 35, 36]. The increase in solubility was due to the increase in polarity as a result of chain scission under irradiation and the decrease in inter-chain hydrogen bonds [35]. Such behavior demonstrates clearly that the starch molecules suffered important changes as a consequence of a degradation phenomenon induced by ionizing radiation processing.

The ionizing radiation processing of all types of starches caused the reduction of **swelling power** as the increase of the irradiation dose, especially at higher doses [5, 9, 10, 15–17, 19, 21, 22, 31, 32, 36–38]. This evolution could be attributed to the fact that starch granules become sensitive being weaker and easier to break after irradiation. In addition, a consequence of starch radiation-induced degradation can be also the inhibition of granule ability to trap water and provoke the swelling explaining thus the reduction of swelling power by irradiation.

3.2. Rheological properties

Ionizing radiation processing is able to produce significant changes in the rheological properties of starch especially by decreasing its viscosity. In this way, the most studies approached the evaluation of pasting behavior of irradiated starches. Thus, a considerable decrease in the paste viscosities was noted as the irradiation dose increased (up to 500 kGy) for starches with different botanical origins [2, 3, 6, 9, 12, 13, 15–17, 19–23, 27, 31–33, 35, 38]. However, exceptions were reported for the breakdown viscosity that increased with the irradiation dose up to 10 kGy for corn or wheat starches [35, 37, 39]. In addition, a comparative study on corn starch treated with the same gamma irradiation dose (10 kGy) in the dose rate range of 0.4–2 kGy/h clearly revealed that the viscosities of the starch pasting profile decreased more at lower dose rates in comparison to native starch [37].

The level of radiation-induced changes in the pasting profile was different according to the starch variety [20, 22, 31] and may be assigned to difference in extent of polymerization of leached amylose and amylopectin molecules of each starch variety. The reduction of the peak viscosity of starch was assigned to its weaker water binding capacity, granular rigidity, and integrity due to glycosidic bond cleavage [35, 40]. Moreover, the decrease in the setback and final viscosities were attributed to the degradation or shortening of amylose and longer amylopectin branch chains by irradiation [17, 37].

A gradual decrease of the initial pasting and peak temperatures was also induced by irradiation [3, 9, 17, 19, 21, 32, 33, 35, 37]. Although the peak time was not influenced by irradiation dose rate, it depended on starch variety [31].

Likewise, the apparent viscosity of irradiated starches decreased significantly as the irradiation dose increased for different cereal and tuber starches [4, 5, 8, 10, 11]. Kamal et al. [8] demonstrated that the electron beam effect on the apparent viscosity of corn starch was greater than that of gamma radiation in the early stage of irradiation ~5 kGy. Moreover, a mathematical model was elaborated to describe the exponential decrease of the apparent viscosity against irradiation dose [4]. At the same time, it was proved that each starch is characterized by a material constant that indicates the functional sensitivity of starch to irradiation. Consequently, in technological applications, based on this model and the material constant typical for each starch, one can calculate the irradiation dose required to be applied in order obtain a certain value of the apparent viscosity.

3.3. Gelatinization

Gelatinization is one of the most important functional properties of starch. Ionizing radiation processing of starch generally leads to great modifications of gelatinization temperatures and process enthalpy due to structural reconfiguration occurring in starch macromolecule. Lately, starch gelatinization is studied and monitored by differential scanning calorimetry (DSC), which is an extremely valuable tool to provide a quantitative measure of the gelatinization enthalpy and a determination of temperature range where gelatinization occurs as well.

From the outset, it should be emphasized that the multitude of experimental data reported in a large volume of papers shows that although the gelatinization properties of starch are affected by irradiation, a pattern of alterations cannot be identified. More specifically, it can be claimed that the evolution of gelatinization temperatures and enthalpy is practically unpredictable for various irradiation conditions and types of starch.

Several investigations [1, 2, 5, 6, 35] reported the decrease of both gelatinization temperatures (onset, peak and conclusion temperatures) and enthalpy as the irradiation increasing (up to 50 kGy) for cereal starches. However, Liu et al. [35] reported that the gelatinization parameters of corn starch almost remained constant under 20 kGy, and afterwards, their significant decrease occurred for irradiation doses up to 500 kGy. Later on, certain decrease in the gelatinization temperatures was reported for corn starch with different amylose content, up to 50 kGy, but only marginal effect on enthalpy values was identified [1]. These results indicated that gamma irradiation caused the production of defective crystalline structure and an increase in the proportion of short chains in amylopectin, which caused a decrease in gelatinization temperature [17]. The decrease in enthalpy value was explained by the disruption of the crystalline domain of starch granules in addition to disruption of double helical order [2, 17].

Other investigations [10] revealed that the gelatinization temperatures and enthalpy had no statistically significant alteration after irradiation with gamma rays for wheat starch treated with irradiation doses up to 50 kGy at a dose rate of 13 Gy/min. More than that, another study [39], using a higher dose rate (1 kGy/h) in the irradiation dose range up to 9 kGy, pointed out no significant difference in gelatinization temperatures and enthalpy for wheat starch after irradiation.

The investigations on rice starch irradiated at low rate of 0.4 kGy/h, with irradiation doses up to 10 kGy [15], also showed no significant shift of gelatinization temperatures up to 5 kGy, confirming the previously reported results [14], but a decrease of gelatinization temperatures and enthalpy was observed after 10 kGy irradiation. Similar results showing no important alteration of gelatinization parameters were also reported for elephant foot yam starch treated with doses up to 25 kGy at a dose rate of 2 kGy/h [19].

An extensive study on four varieties of starch extracted from the beans [20] revealed the reduction of the gelatinization temperatures and enthalpy of bean starch by irradiation with doses up to 25 kGy at low dose rate of 185 Gy/h. Contrary, Chung et al. [17] have found that the gelatinization temperatures for bean starch remained unaffected at 10 kGy and increased slightly at 50 kGy (2 kGy/h). An increasing behavior of gelatinization temperatures has also been reported for potato starch exposed to e-beam up to 50 kGy at high dose rate (2 kGy/min) [5], while the gamma irradiation at a dose rate of 2 kGy/h caused the increase of gelatinization temperatures when irradiated at 10 kGy, but decreased at 50 kGy [17]. A significant increase in the onset and peak temperatures was reported while no important effect on the gelatinization enthalpy was noticed for sago starches under irradiation treatment with doses less than 25 kGy [36]. Increase in gelatinization temperatures in irradiated starches was correlated with decreases in the overall crystallinity resulting that among the starch crystallites containing various rigidities, the relatively weak crystalline structure could be preferentially destroyed during irradiation [17].

4. Effects of ionizing radiation on structure

The investigation of ionizing radiation effects on starch structure revealed information related to the granule morphology, crystalline structure, or structural characteristics as determined by X-ray diffraction (XRD), scanning electron microscopy (SEM), Fourier transform infrared (FTIR) spectroscopy and chromatography.

4.1. Crystallinity

Crystallinity and crystallinity degree of starch macromolecule can be evaluated by different analytical techniques of investigation, such as X-ray diffraction, infrared spectroscopy, or DSC. The same starch granule consists of both crystalline regions (crystalline lamellae of amylopectin) and amorphous regions (typical to amylose) without a net delimitation making the determination of its crystallinity actually difficult.

One of the most used method to analyze the crystallinity and crystallinity degree is the X-ray diffraction because it can provide the crystallographic patterns of starch granules. The crystalline lamellae show two types of polymorph structures that design different diffraction patterns: the A-type crystallinity with relatively compact structure, the B-type crystallinity with a more open structure, including a hydrated helical core, and the C-type crystallinity that is a mixture of A-type and B-type patterns [41]. The crystallinity and crystallinity degree depend on the botanical source of starch and amylose content [1] or distribution of chain length in amylose. The experimental data proved that the diffraction pattern remained generally unaffected even at very high irradiation doses up to 500 kGy [35]. However, Reddy et al. [19] reported alteration of crystallographic pattern of elephant foot yam starch by gamma radiation processing. Thus, the B-type pattern of native starch changed in the C-type in irradiated starch.

On the other hand, most investigations recorded the decrease in the crystallinity degree of starch as a result of irradiation. The reduction in crystallinity degree with the increasing irradiation dose has been explained by breaking of the crystalline regions. Conversely, some studies [2, 10] found out that the degree of crystallinity was insignificantly changed for cereal starches by gamma irradiation processing with irradiation doses up to 50 kGy and low irradiation dose rates (<19 Gy/min). In case of wheat starch, Kong et al. [39] reported that the gamma irradiation at a dose rate of 1 kGy/h moderately affected the crystallinity degree that increased continuously with irradiation dose increasing up to 7 kGy and decreased at 9 kGy. This behavior was attributed to the alterations predominantly in amorphous regions induced with irradiation dose up to 7 kGy, whereas the crystalline regions were more affected by irradiation dose of 9 kGy. Also, the irradiation at 10 kGy with dose rate of 0.4 kGy/h caused an increase of the crystallinity degree of corn starch indicating that a slower dose rate is able to induce more crystalline structure [37]. Moreover, the investigation on rice starch [14] revealed both reduction and an increase in crystallinity degree by irradiation up to 5 kGy with dose rate of 0.4 kGy/h depending on rice cultivars. Therefore, the ionizing radiation processing influenced both the crystalline region and amorphous region of starch granules, leading to the decrease or increase in crystallinity degree in accordance with the most affected region by irradiation dose. Anyway, the investigators considered even the possibility of crosslinking in the case of the increase of crystallinity degree.

Another study [17] stated that the crystallinity degree decreased more rapidly with irradiation dose increasing for bean starch (C-type pattern) than potato starch (A-type pattern), showing that the crystallographic patterns have different sensitivity to irradiation. Thus, the B-type pattern has been proven to be more sensitive to ionizing radiation processing than A-type pattern and justified the behavior of bean starch in which the B-type pattern degraded faster than A-type pattern from granule surface. It is noteworthy that recently an opposite behavior has been identified by Chung et al. [1] for corn starch with different content of amylose subjected to ionizing radiation processing, namely the high amylose corn starch showed a B-type pattern being more radiation resistant than the waxy corn starch with A-type pattern.

Several works [3, 10, 17, 21, 33, 37] have reported results on the crystallinity degree of irradiated starch, estimated by using infrared spectroscopy, which involves the analysis of the absorption bands at 1047 cm^{-1} (crystalline structure) and 1022 cm^{-1} (amorphous region), and their respective ratio indicates the degree of starch order. The experimental data revealed that the ratio of 1047/1022 decreased with an increase of gamma radiation dose up to 50 kGy, and the starch granular crystallinity was affected [17, 21, 33, 37]. However, a couple of studies [3, 10] showed that the ionizing radiation processing (up to 50 kGy) had no influence on this ratio, suggesting that larger crystalline regions might be broken into small crystallites such that the crystallinity degree was practically unaffected.

4.2. Granule morphology

Morphology of starch granule exposed to ionizing radiation can be affected depending on starch type and irradiation parameters. For instance, studies on potato and rice starches [16, 38] reported the surface cracking of granules as well as deformation of granular structure were identified to increase with increasing irradiation dose in the range of 5–20 kGy at a dose rate of 2 kGy/h. Also, the extent of change depended on starch variety. Other studies on rice starches [14, 15] at lower irradiation doses (<10 kGy) and lower dose rate (0.4 kGy/h) also showed some modifications in the values of the mean sizes of the granules depending on irradiation dose and rice cultivars even if the irradiation apparently caused no change in the granule morphology. On the contrary, Shishonok et al. [18] reported that the surface structure of potato starch suffered no damage by electron beam irradiation even at high irradiation doses (110–440 kGy).

For corn starch irradiated with gamma rays up to 50 kGy and dose rate around 1 kGy/h, an absence of notable changes on the shapes and sizes of starch granules has been noticed [1, 2]. These observations were confirmed and completed by another investigation [35], which reported that gamma irradiated corn starch retained the original shape and size without any granular cracking or roughness occurring on the surface, even for 500 kGy with a dose rate 83 Gy/min. On the other hand, other investigations found changes in corn starch morphology induced by ionizing radiation. Although the granule shape and sizes were apparently unaffected by electron beam processing, the appearance of small circular perforations on the granule surface could be observed for irradiation of 50 kGy (dose rate of 2 kGy/min) [3]. Moreover, Kamal et al. [8] showed that the shape of corn starch granule was somewhat deformed by both gamma rays and electron beams for doses up to 100 kGy. It is noteworthy that the different content of amylose in corn starch had no influence on morphological aspects of irradiated starch; the granules remained intact and visually unchanged by gamma irradiation up to 50 kGy with dose rate around 1 kGy/h [1].

Microscopic observation of bean starches indicated surface cracking of granule with irradiation in a dose-dependent manner in the irradiation dose range of 5–25 kGy (<185 Gy/h) without significant changes in granule dimensions [20, 21]. However, in an earlier study, Gani et al. [22] have found the deformation of granule increased with increasing irradiation dose in the range of 5–20 kGy (dose rate of 2 kGy/h), the extent of change depending on starch variety. Other starches extracted from different botanical sources (lotus, chickpea) also presented surface fissures induced by irradiation [23, 32], while the ionizing radiation processing had no influence on the morphological characteristics for starches from elephant foot yam, Indian horse chestnut, and sago [19, 33, 36].

Consequently, the ionizing radiation is an energetic penetrating radiation that able to produce effects in the whole volume of the samples, so that the radiation-induced changes may occur both in the central regions and in the peripheral regions of the starch granules. However, the fact that microscopic methods reveal no damage to the granule outer layer for some starches leads to the conclusion that the radiation-induced changes might occur at a more intimate level of matter in the form of structural changes depending on starch granular structure.

4.3. Spectral characteristics

Generally, the analytical evaluation of FTIR spectrum of native starch must show five different frequency regions as presented in **Table 3** [42–44]. Therefore, potential modifications induced by ionizing radiation processing of bands assigned to those frequency regions should be evaluated.

The spectral features of the irradiated starch were apparently similar to native starch and no bands of new functional groups were found in spectrograms [3, 9, 10, 15, 18, 19, 33, 35]. For instance, Liu et al. [35] found that all spectral patterns for corn starch irradiated with gamma radiation were similar to those of control sample even after 500 kGy irradiation. However, some differences related to the frequency and intensity of some bands were identified after irradiation indicating radiation-induced alteration of the macromolecule structural integrity. Thus, slight shifts of some peaks and the decrease in intensity of some bands or the increase in intensity of other bands with the irradiation dose increasing have been noticed for all types of starches. The most affected bands were especially those assigned to O—H and C—H bonds [2, 3, 8], indicating that the stability of the inter- and intramolecular hydrogen bonds of starch structure was affected by ionizing radiation processing [3]. As an example, Bettaïeb et al. [2] found intensity decrease with 38.4 and 19.6%, respectively, for corn starch after irradiation with 50 kGy at 19 Gy/min. Instead, for wheat starch [10], the absorbance intensity of the same bonds decreased dramatically about 70 and 67%, respectively, after irradiation with 50 kGy at

http://dx.doi.org/10.5772/intechopen.71626

Frequency region [cm^{-1}]	Assignment
3000–3700	O–H stretch
2800–3000	C–H stretch
1550–1800	O–H vibrations from bound water molecules
800–1550	Fingerprint region
Below 800	Pyranose ring of the glycosidic unit

Table 3. Band region frequencies and assignments of FTIR absorption of starch.

13.84 Gy/min. These results were attributed to the breaking of chemical bonds by irradiation. Besides, the botanical source of starch and the irradiation dose rate influenced the degree of the radiation-induced changes. Also, the findings [2] showed a decrease in peak intensity of the bending mode of the glycosidic linkage (C–O–C) with 13.4% explained by a depolymerization of amylose chains of starch and/or the amylopectin double helices within the amorphous regions after irradiation due to the breaking of glycosidic linkages [2, 10]. Conversely, an increase in the intensity of the characteristic peak at 1647 cm^{-1} ascribed to carbonyl groups was also observed [8] for e-beam irradiated corn starch, suggesting that the starch degraded by free radical reaction.

At the same time, the band intensity of the bending mode of water was also affected by a decreasing trend as the irradiation dose increased [2, 10]. This change occurred by water radiolysis that involves the breakdown of the water structure under the action of ionizing energy and leads to the formation of hydroxyl and hydrogen radicals.

4.4. Molecular weight and molecular weight distribution

Molecular weight of polymers influences most of their physicochemical and functional properties, and its investigation can reveal information useful to understand the behavior of macromolecules to ionizing radiation processing. Unfortunately, one can notice the lack of interest in this subject and the existence of only a few papers [4, 5, 12, 17] that have approached the study of the influence of the electron beam or gamma radiation on the starch molecular weights. The experimental data showed the decreasing evolution of the molecular weights with the irradiation dose independently of the starch botanical source. This kind of behavior indicated the break of polymeric chain and formation of the fragments with different molecular weights, which modified the mass molecular distribution of starch. However, the radiation-induced changes were correlated with the structural organization of starch, especially the branched structure component of starch (amylopectin). Hence, for cereal starches having short chains, the molecular weight distribution was affected mainly by the formation of the fractions with higher molecular weight than the formation of the fractions with low molecular weight. Instead, for tuber starch having long chains, the molecular weight distribution was slightly modified by irradiation, namely the scissions in fractions with high molecular weight being closer to that of the fractions with low molecular weight.

5. Concluding remarks and future perspectives

The great volume of results published in the last decade indubitably proved that the ionizing radiation (gamma rays and electron beam) is able to produce changes in structural and functional properties of starch, mainly due to the degradation process.

The radiation-induced effects are related to depolymerization of starch macromolecule followed by the reduction of molecular weight as well as the alteration of the double helix in the branched regions and crystalline structure, especially in the intrusion area of the amorphous region in the crystalline structure. Consequently, irradiation induces generally reconfiguration of starch molecules which lead to the reduction of crystallinity degree, shifts and decrease of spectral bands, and changes of thermal parameters. The general trend of decreasing for viscosity and swelling power concomitantly with the increasing of water solubility by irradiation makes the irradiated starches able to meet the specific needs in different new applications.

It is important to note that a number of factors related both to starch and ionizing radiation processing plays a major role in dictating the response of granular starch to ionizing radiation (**Table 4**). Taking into account this aspect, the comparison of properties among the irradiated starches should attentively be performed due to the differences in methodologies of ionizing radiation treatments and in composition and structure of starches. In other words, the random approaches of ionizing radiation processing of starch extracted from various botanical sources can lead to various results making difficult their comparison and the identification of a typical pattern behavior of a specific starch property.

Further studies should systematically focus on the response (physicochemical and structural properties) of each type of starch having different moisture content exposed to ionizing radiation (gamma rays and/or electron beam) in a large range of irradiation dose rate and different gas atmospheres. Studies on the major starch components, amylose and amylopectin, extracted from different native starches and subsequently exposed to ionizing radiation can be useful to validate observations on starch, leading to advancements in this research area.

Another issue that must be carefully explored is related to the investigation of thermal properties by DSC since nowadays the available reports showed a large variability of results without consistent correlations with other structural investigated properties of starch.

It is also opportune to make deeper chromatographic studies on the molecular weight and mass distribution of irradiated starch, especially as the chromatographic technique has developed spectacularly in recent years. The comprehensive evaluation of the dynamics of molecular mass distribution of irradiated starch will provide new relevant knowledge, contributing to a better

Starch factors	Ionizing radiation processing
Type of starch (botanical source)	Irradiation dose
Variety of starch (cultivar)	Dose rate
Water content	Type of gas atmosphere

Table 4. Factors influencing the response of starch to ionizing radiation processing.

http://dx.doi.org/10.5772/intechopen.71626

understanding and even to behavior prediction of specific functional characteristics of starches exposed to ionizing radiation processing.

Acknowledgements

This work was supported by project NUCLEU.

Author details

Mirela Brașoveanu and Monica-Roxana Nemțanu*

*Address all correspondence to: monica.nemtanu@inflpr.ro

National Institute for Lasers, Plasma and Radiation Physics, Electron Accelerators Laboratory, Bucharest-Măgurele, Romania

References

[1] Chung KH, Othman Z, Lee JS. Gamma irradiation of corn starches with different amylose-to-amylopectin ratio. Journal of Food Science and Technology. 2015;**52**(10):6218-6229

[2] Bettaïeb NB, Mohamed TJ, Ghorbel D. Gamma radiation influences pasting, thermal and structural properties of corn starch. Radiation Physics and Chemistry. 2014; **103**:1-8

[3] Braşoveanu M, Nemţanu MR, Duţă D. Electron-beam processed corn starch: Evaluation of physicochemical and structural properties and technical-economic aspects of the processing. Brazilian Journal of Chemical Engineering. 2013;**30**(4):847-856

[4] Nemtanu MR, Brasoveanu M. Radio-sensitivity of some starches treated with accelerated electron beam. Starch-Starke. 2012;**64**:435-440

[5] Nemţanu MR, Braşoveanu M. Chapter 17: Functional properties of some non-conventional treated starches. In: Eknashar M, editor. Biopolymers. Rijeka, Croatia: Scyio; 2010. p. 319-344

[6] Nemtanu MR, Minea R, Kahraman K, Koksel H, Ng PKW, Popescu MI, Mitru E. Electron beam technology for modifying the functional properties of maize starch. Nuclear Instruments and Methods in Physics Research A. 2007;**580**(1):795-798

[7] Nemtanu MR. Influence of the electron beam irradiation on the colorimetric attributes of starches. Romanian Journal of Physics. 2008;**53**(7-8):873-879

[8] Kamal H, Sabry GM, Lotfy S, Abdallah NM, Ulanski P, Rosiak J, Hegazy E-SA. Controlling of degradation effects in radiation processing of starch. Journal of Macromolecular Science, Part A. 2007;**44**:865-875

[9] Bashir K, Swer TL, Prakash KS, Aggarwal M. Physico-chemical and functional properties of gamma irradiated whole wheat flour and starch. LWT- Food Science and Technology. 2017;**76**:131-139

[10] Atrous H, Benbettaieb N, Hosni F, Danthine S, Blecker C, Attia H, Ghorbel D. Effect of γ-radiation on free radicals formation, structural changes and functional properties of wheat starch. International Journal of Biological Macromolecules. 2015;**80**:64-76

[11] Nemtanu MR, Brasoveanu M. Aspects regarding the rheological behavior of wheat starch treated with accelerated electron beam. Romanian Journal of Physics. 2010;**55**(1-2):111-117

[12] Hu B, Huang M, Yin S, Zi M, Wen Q. Effects of electron-beam irradiation on physicochemical properties of starches separated from stored wheat. Starch-Starke. 2011;**63**:121-127

[13] Adzahan NM, Hashim DM, Muhammad KR, Rahman A, Ghazali Z, Hashim K. Pasting and leaching properties of irradiated starches from various botanical sources. International Food Research Journal. 2009;**16**:415-429

[14] Polesi LF, Sarmento SBS, de Moraes J, Franco CML, Canniatti-Brazaca SG. Physico-chemical and structural characteristics of rice starch modified by irradiation. Food Chemistry 2016;**191**:59-66

[15] Gul K, Singh AK, Sonkawade RG. Physicochemical, thermal and pasting characteristics of gamma irradiated rice starches. International Journal of Biological Macromolecules. 2016;**85**:460-466

[16] Gani A, Nazia S, Rather SA, Wani SM, Shah A, Bashir M, et al. Effect of γ-irradiation on granule structure and physicochemical properties of starch extracted from two types of potatoes grown in Jammu & Kashmir, India. LWT- Food Science and Technology. 2014;**58**:239-246

[17] Chung H-J, Liu Q. Molecular structure and physicochemical properties of potato and bean starches as affected by gamma-irradiation. International Journal of Biological Macromolecules. 2010;**47**:214-222

[18] Shishonok MV, Litvyak VV, Murshko EA, Grinyuk EV, Sal'nikov LI, Roginets LP, et al. Structure and properties of electron beam irradiated potato starch. High Energy Chemistry. 2007;**41**:425-429

[19] Reddy CK, Suriya M, Vidya PV, Vijina K, Haripriya S. Effect of γ-irradiation on structure and physico-chemical properties of *Amorphophallus paeoniifolius* starch. International Journal of Biological Macromolecules. 2015;**79**:309-315

[20] Hussain PR, Wani IA, Suradkar PP, Dar MA. Gamma irradiation induced modification of bean polysaccharides: Impact on physicochemical, morphological and antioxidant properties. Carbohydrate Polymers. 2014;**110**:183-194

[21] Sofi BA, Wani IA, Masoodi FA, Saba I, Muzaffar S. Effect of gamma irradiation on physicochemical properties of broad bean (Vicia faba L.) starch. LWT- Food Science and Technology. 2013;**54**:63-72

[22] Gani A, Bashir M, Wani SM, Masoodi FA. Modification of bean starch by γ-irradiation: Effect on functional and morphological properties. LWT- Food Science and Technology. 2012;**49**:162-169

[23] Gani A, Gazanfar T, Jan R, Wani SM, Masoodi FA. Effect of gamma irradiation on the physicochemical and morphological properties of starch extracted from lotus stem harvested from Dal lake of Jammu and Kashmir, India. Journal of the Saudi Society of Agricultural Sciences. 2013;**12**:109-115

[24] Bhat R, Karim AA. Impact of radiation processing on starch. Comprehensive Reviews in Food Science and Food Safety. 2009;**8**:44-58

[25] Zhu F. Impact of γ-irradiation on structure, physicochemical properties, and applications of starch. Food Hydrocolloids. 2016;**52**:201-212

[26] Nemţanu MR, Braşoveanu M. Chapter 5: Electron beam treatment as a non-conventional preservation method of nutritional supplements based on vegetal materials. In: Hulsen I, Ohnesorge E, editors. Food Science Research and Technology. New York: Nova Science Publishers, Inc.; 2009. p. 169-192

[27] Ghisleni DDM, Braga MS, Kikuchi IS, Braşoveanu M, Nemţanu MR, Dua K, et al. The microbial quality and decontamination approaches for the herbal medicinal plants and products: An in-depth review. Current Pharmaceutical Design. 2016;**22**(27):4264-4287

[28] ICRU. Radiation quantities and units. Report 33. Washington; 1980

[29] IAEA. Dosimetry for food irradiation. Technical Reports Series No. 409. Vienna; 2002

[30] IAEA. Absorbed dose determination in photon and electron beams. Technical Reports Series No. 277. Vienna; 1987

[31] Ocloo FCK, Owureku-Asare M, Agyei-Amponsah JW, Agbemavor SK, Egblewogbe MNYH, Apea-Bah FB, et al. Effect of gamma irradiation on physicochemical,functional and pasting properties of some locally-produced rice (*Oryza* spp) cultivars in Ghana. Radiation Physics and Chemistry. 2017;**130**:196-201

[32] Bashir M, Haripriya S. Physicochemical and structural evaluation of alkali extracted chickpea starch as affected by γ-irradiation. International Journal of Biological Macromolecules. 2016;**89**:279-286

[33] Wani IA, Jabeen M, Geelani H, Masoodi FA, Saba I, Muzaffar S. Effect of gamma irradiation on physicochemical properties of Indian horse chestnut (*Aesculus indica* Colebr.) starch. Food Hydrocolloids. 2014;**25**:253-263

[34] Ershov BG. Radiation-chemical degradation of cellulose and other polysaccharides. Russian Chemical Reviews. 1998;**67**:315-334

[35] Liu T, Ma Y, Xue S, Shi J. Modifications of structure and physicochemical properties of maize starch by γ-irradiation treatments. LWT- Food Science and Technology. 2012;**46**:156-163

[36] Othman Z, Hassan O, Hashim K. Physicochemical and thermal properties of gamma-irradiated sago (Metroxylon Sagu) starch. Radiation Physics and Chemistry. 2015;**109**:48-53

[37] Chung H-J, Liu Q. Effect of gamma irradiation on molecular structure and physicochemical properties of corn starch. Journal of Food Science. 2009;**74**:C353-C361

[38] Ashwar BA, Shah A, Gani A, Rather SA, Wani SM, Wani IA, et al. Effect of gamma irradiation on the physicochemical properties of alkali-extracted rice starch. Radiation Physics and Chemistry. 2014;**99**:37-44

[39] Kong X, Zhou X, Sui Z, Bao J. Effects of gamma irradiation on physicochemical properties of native and acetylated wheat starches. International Journal of Biological Macromolecules. 2016;**91**:1141-1150

[40] Lee YJ, Kim SY, Lim ST, Han SM, Kim HM, Kang IJ. Physicochemical properties of gamma-irradiated corn starch. Journal of Food Science and Nutrition. 2006;**11**:146-154

[41] Tester RF, Karkalas J, Qi X. Starch—Composition, fine structure and architecture. Journal of Cereal Science. 2004;**39**:151-165

[42] Nemţanu MR, Braşoveanu M. Degradation of amylose by ionizing radiation processing. Starch-Starke. 2017;**69**(3-4):1600027. DOI: 10.1002/star.201600027

[43] Kizil R, Irudayaraj J, Seetharaman KJ. Characterization of irradiated starches by using FT-Raman and FTIR spectroscopy. Journal of Agricultural and Food Chemistry. 2002; **50**:3912-3918

[44] Galat A. Study of the Raman scattering and infrared absorption spectra of branched polysaccharides. Acta Biochimica Polonica. 1980;**27**:135-142